RF Plasma Heating in Toroidal Fusion Devices

RF Plasma Heating in Toroidal Fusion Devices

V. E. Golant and V. I. Fedorov

A. F. Ioffe Physicotechnical Institute
Leningrad, USSR

Translated from Russian by

Donald H. McNeill

Plasma Physics Laboratory
Princeton, New Jersey

CONSULTANTS BUREAU
NEW YORK AND LONDON

Library of Congress Cataloging in Publication Data

Golant, Viktor Evgen'evich.
 RF plasma heating in toroidal fusion devices.

 Translation of: Vysokochastotnye metody nagreva plazmy v toroidal'nykh
termoiadernykh ustanovkakh.
 Bibliography: p.
 Includes index.
 1. Plasma heating. 2. Controlled fusion. 3. Fusion reactors. I. Fedorov, V. I.
(Vladimir Il'ich) II. Title.
QC718.5.H5G6513 1989   621.48′4   88-34198
ISBN-13: 978-1-4684-1673-2   e-ISBN-13: 978-1-4684-1671-8
DOI: 10.1007/978-1-4684-1671-8

This volume is published under an agreement with the
Copyright Agency of the USSR (VAAP)

©1989 Consultants Bureau, New York
Softcover reprint of the hardcover 1st edition 1989

A Division of Plenum Publishing Corporation
233 Spring Street, New York, N.Y. 10013

PREFACE

Because of recent progress in the development of quasistationary toroidal magnetic confinement systems, especially tokamaks, these systems are at the center of research on controlled thermonuclear fusion. Tokamaks were proposed and first built at the Kurchatov Institute of Atomic Energy. In the 1960s the basic features of plasma behavior in toroidal magnetic confinement systems were investigated in experiments on the first tokamaks and the possibility of obtaining effective confinement in them was demonstrated. The successes of this first stage led to a rapid expansion in tokamak research around the world. The development of a thermonuclear power reactor based on the tokamak is now actively under way.

During the earliest phase of research on tokamaks, it was already clear that the ohmic heating used in them was not sufficient to obtain the temperatures needed for initiation of a self-sustaining thermonuclear reaction. At the beginning of the 1970s, therefore, a search was begun for methods of heating which could supplement ohmic heating. The best of these auxiliary heating techniques are neutral beam injection, various methods based on the collisionless absorption of rf (radio frequency) waves, and adiabatic compression of the plasma by a rising magnetic field. The greatest successes have been achieved in the development and application of neutral beam heating, with which temperatures close to those needed for controlled thermonuclear fusion have been obtained in modern tokamaks. Major technical difficulties, however, stand in the way of neutral beam heating for a thermonuclear reactor. Thus, despite the successes of neutral beam injection, the development of rf methods for plasma heating continues apace.

During the last ten years a very large volume of theoretical and experimental research has been done on rf heating of tokamak plasmas, along with a whole series of associated engineering developments. Most of these studies have been done in four frequency bands where the waves undergo collisionless absorption at different types of resonances in the plasma. These bands correspond to the electron cyclotron, ion cyclotron, lower hybrid, and Alfvén resonance frequencies. Along with these physical studies, experiments have recently been carried out on the heating of plasmas in tokamaks with rf input powers considerably higher than the ohmic heating power. In the ion cyclotron frequency range the rf power has been as high as 4 MW, while in the electron cyclotron and lower hybrid frequency ranges, heating powers of up to 1 MW have been used. The resulting peak temperatures were considerably higher than those obtained with ohmic heating and approach those achieved with neutral beam injection. More extensive experiments on rf heating in stellarators are also under way.

Hundreds of articles on rf plasma heating have been published in recent years. Special conferences, meetings, and symposia on heating techniques have been organized. Of these, a special role is played by the biennial international symposia on heating in toroidal plasmas which are held alternately in France and Italy. Since almost all the principal results of research on rf heating have been presented at these symposia, their proceedings [1–4] serve as a unique chronicle of progress in rf plasma heating in toroidal machines. Review articles on rf heating have also been published in the past few years. Comparatively short reviews have recently been published by Alikaev [5] and Suvorov [6]. More detailed reviews devoted to various methods of rf heating and the phenomena associated with them have been collected in the book *Rf Plasma Heating* prepared by the Institute of Applied Physics of the Academy of Sciences of the USSR [7]. It contains reviews by Alikaev, Litvak, Suvorov, and Fraiman on electron cyclotron heating, Longinov and Stepanov on ion cyclotron heating, and Elfimov, Kirov, and Sidorov on Alfvén wave heating.

The purpose of the present book is to provide, in seven chapters, a unified overview of the methods for rf heating of plasmas in toroidal fusion experiments. In Chapter 1 the problem of plasma heating in tokamaks and stellarators is formulated and the requirements for auxiliary heating techniques are described. This chapter also contains a brief review of the results of research on tokamaks and stellarators. Chapter 2 is devoted to a theoretical description of the principal physical effects involved in the rf heating of plasmas, especially the characteristics of wave propagation, of the mechanisms by which waves are absorbed and plasma heating takes place, and of the nonlinear effects that accompany heating. The primary emphasis is on a qualitative physical picture of these effects. Chapters 3–6, in turn, deal with the major rf heating techniques currently under investigation, electron cyclotron (ECH), ion cyclotron (ICH), lower hybrid (LHH), and Alfvén wave heating. In each of these chapters the main schemes for heating are described, the results of theoretical analyses and numerical simulations are discussed, the technology of the heating systems is briefly described, and experimental work published through the end of 1984 is reviewed. Finally, in Chapter 7 the different rf heating techniques are compared; they are contrasted with neutral beam injection, and the feasibility of adiabatic compression as a means of heating plasmas is examined. The list of references does not claim to be complete. Only the basic papers on rf plasma heating in tokamaks and stellarators have been included. We have used standard notation for the main variables. In order to avoid confusion, we note that in the text a script Latin v is used for the velocity, which is close in form to the Greek ν used for the collision frequency.

This book is intended primarily for scientists, engineers, and students who are beginning to work on or have an interest in rf plasma heating.

CONTENTS

Chapter 1

PLASMA HEATING IN
TOROIDAL FUSION DEVICES

1.1. CONDITIONS FOR ENERGY PRODUCTION
IN QUASISTATIONARY SYSTEMS

The thermonuclear power reactors currently under discussion are based on the fusion of deuterium and tritium nuclei [8–10]:

$$D + T \rightarrow {}^4He \ (3.52 \ \text{MeV}) + n \ (14.06 \ \text{MeV}).$$

The fuel in such a reactor must be deuterium. It is proposed that tritium be recovered in a lithium blanket through the reaction

$$ {}^6Li + n \rightarrow T + {}^4He + 4.8 \ \text{MeV}.$$

The power p_F released per unit volume of the reactor by thermonuclear reactions is determined by the density of reacting nuclei n, their temperature T, and the reaction cross section σ. For a 50–50 mixture of deuterium and tritium,

$$p_F = (1/4) \, n^2 \langle \sigma v \rangle q_F, \qquad\qquad (1.1)$$

where v is the relative speed of the nuclei; $\langle \sigma v \rangle$ is averaged over a Maxwellian velocity distribution and depends on the temperature; and $q_F = 22.4$ MeV is the energy released by the (D, T) reaction and the subsequent (Li, n) reaction.

A criterion for positive energy yield may be obtained by comparing the power released in a volume with the energy lost from the volume per unit time through radiation and by transport. The specific loss power p_L is usually characterized by the total energy confinement time, i.e.,

$$p_L = W/\tau_E = 3\overline{nT}/\tau_E,\qquad(1.2)$$

where W is the stored energy per unit volume of the plasma and T is the plasma temperature. (The temperatures of the electron and ion components are assumed to be equal.)

In order for the energy released in the reactor (after it has been converted into plasma heating energy) to exceed the lost energy, the condition

$$\eta\,(p_F + p_L) > p_L$$

or

$$n\tau_E > \frac{12\,(1-\eta)}{\eta}\,\frac{T}{q_F\langle\sigma v\rangle}\qquad(1.3)$$

must be satisfied. Here η is the conversion efficiency. The inequality (1.3), known as the Lawson criterion, establishes the dependence of $(n\tau_E)_{min}$ on T. This dependence is shown in Fig. 1.1 (curve 1) for the customarily assumed value of $\eta = 1/3$.[*] It is clear that for a temperature above 10 keV a positive energy yield is possible when $n\tau > 10^{14}$ cm^{-3}·s. In a steady-state or quasi-steady-state reactor, the most convenient operating regime is one in which the required temperature is maintained by the energy of the α particles formed in the (D, T) reaction. Evidently, the condition for such a self-sustaining reaction is that the energy of the α particles formed per unit time must exceed the corresponding energy loss from the plasma:

$$p_\alpha = (1/4)n^2\langle\sigma v\rangle q_\alpha > p_L$$

or

$$n\tau_E > 12\,T/(q_\alpha\langle\sigma v\rangle),\qquad(1.4)$$

where $q_\alpha = 3.52$ MeV is the energy of an α particle formed in the (D, T) reaction. This inequality is referred to as the condition for ignition of a thermonuclear reaction. The dependence of $(n\tau_E)_{min}$ on T corresponding to this condition is shown in Fig. 1.1 by curve 2. At $T \sim 10$ keV the value of $n\tau_E$ required for ignition is $3 \cdot 10^{14}$ cm^{-3}·s.

These criteria should be supplemented by a restriction that follows from the requirement that a reactor should produce a sufficiently high specific power. For economic reasons, it is customarily assumed that $p_F > 1$ W/cm^3. This condition, together with Eq. (1.1), places a limit on the plasma density:

---

[*]The usual form of the Lawson criterion differs from Eq. (1.3) in that losses through electron bremsstrahlung are included, but this is important only at low temperatures (below 7 keV) where bremsstrahlung forms a significant part of the total.

$$n > (4p_{min}/q_F \langle \sigma v \rangle)^{1/2}. \qquad (1.5)$$

The dependence of n_{min} on T is shown in Fig. 1.2. For $T \sim 10$ keV, $n_{min} \sim 10^{14}$ cm^{-3}. For steady-state and quasistationary systems the plasma density cannot be much greater than this minimum. This limitation exists primarily because an increase in the density leads to a rapid growth in the thermal load on the reactor walls. The energy released per unit time per unit wall area is given by

$$Q = (\overline{p}_F + \overline{p}_L) V/S,$$

where V is the plasma volume and S is the surface area of the walls. The practical limit on Q is usually taken to be on the order of 200 W/cm^2. Since $V/S \sim 50$ for the reactors currently under discussion, this implies that $p_F < QS/V \sim 4$ W/cm^3 or

$$n < \left(\frac{4QS}{q_F \langle \sigma v \rangle V} \right)^{1/2} \simeq \left(\frac{16}{q_F \langle \sigma v \rangle} \right)^{1/2}. \qquad (1.6)$$

The value n_{max} corresponding to this inequality (Fig. 1.2) is only twice n_{min} given by Eq. (1.5). The resulting density range determines the required energy confinement time for the plasma in a quasistationary thermonuclear reactor. For self-sustained operation, these densities follow from the ignition condition (1.4), which limits $(n\tau_E)_{min}$, and the inequality (1.6), which limits n (Fig. 1.2). For $T = 10$ keV the confinement must obey $\tau_E > 2$ s.

For many years, research on magnetic confinement has been aimed at achieving the confinement required for quasistationary systems. The greatest success has been obtained in closed (toroidal) magnetic confinement systems, primarily tokamaks. The values of $n\tau_E$ reached in tokamaks are close to those needed in a thermonuclear reactor. The characteristic feature of toroidal magnetic confinement systems is a relatively weak dependence of the energy confinement time on the plasma temperature. The simplest way of achieving the conditions for energetically favorable thermonuclear fusion can then be represented schematically in two stages: producing a plasma with the required density but a relatively low temperature, and then heating it to the ignition temperature.

In the conventional scheme, therefore, heating the plasma to thermonuclear temperatures is necessary only for bringing the reactor to ignition. In a tokamak the plasma is formed and initially heated by the current flowing in the plasma. This current, however, is limited by the condition for stability and, as an analysis shows, is not sufficient. The maximum temperature obtainable by ohmic heating in a tokamak is 1–2 keV. For ignition ($T \sim 10$ keV) to be reached, other methods (referred to as auxiliary heating) must be applied in addition to ohmic heating. These methods include rf heating, neutral beam heating, and adiabatic compression. They can also be used for plasma heating in currentless toroidal magnetic systems (stellarators).

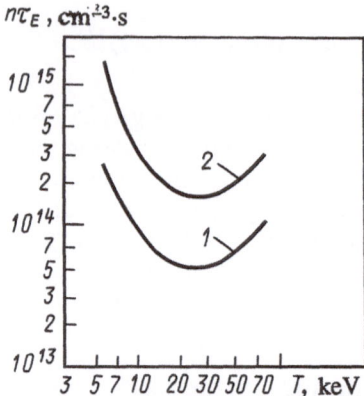

Fig. 1.1. The temperature dependence of $n\tau_E$: 1) the minimum value of $n\tau_E$ for $\eta = 0.33$; 2) $n\tau_E$ corresponding to ignition.

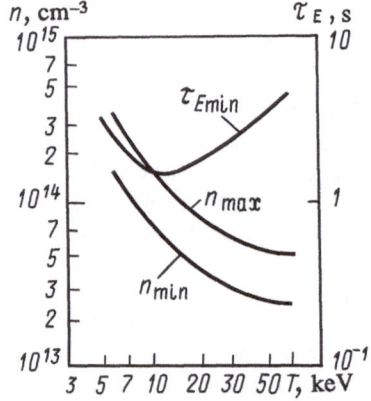

Fig. 1.2. The density limits and required energy confinement time for a thermonuclear reactor.

1.2. BASIC CONCEPTS OF PLASMA CONFINEMENT IN TOKAMAKS

The main concepts of plasma confinement in closed magnetic systems have been developed during the past 25–30 years of theoretical and experimental research. This research has been reviewed in detail elsewhere [11–14]. In closed systems a strong toroidal magnetic field B_φ, whose lines of force form a circle with its center on the axis of the torus, is used to confine the plasma. This field alone, however, is not sufficient to confine the particles since the radial inhomogeneity in the field causes particle drift. The drift of electrons and ions in opposite directions along the axis of the torus causes an increasing charge separation. The resulting electric field, in turn, makes the entire plasma drift in the direction of lower magnetic field. This drift (known as toroidal drift) leads to rapid loss of charged particles on the outer wall. In order to prevent toroidal drift, the toroidal field in closed magnetic confinement systems is combined with a poloidal field B_ϑ created by the currents flowing in the toroidal direction. The superposition of these magnetic field components yields helical field lines on nested toroidal surfaces. Then the field lines, which are not closed, fill the magnetic surfaces entirely so that the field structure in closed systems is composed of a system of nested toroidal surfaces. Because of this field structure, the charged particle drifts associated with the field inhomogeneity are combined with motion along the helical field lines, and the combined trajectory does not extend far from the magnetic surfaces. In addition, the high conductivity of the hot plasma along the magnetic field lines shorts out the electric field caused by charge separation, sharply reducing its strength and the toroidal drift velocity of the plasma.

Two types of toroidal magnetic systems have undergone extensive development: tokamaks, in which the toroidal field is created by external windings and the poloidal field is created by the plasma current; and stellarators, in which both field components are produced by external windings. In a tokamak the toroidal magnetic field (Fig. 1.3) is created by a toroidal solenoid. It is inversely proportional to the major radius of the torus within the cross section of the solenoid. The poloidal field in a tokamak is produced by the toroidal current flowing in the plasma. Its distribution over the plasma cross section depends on the current distribution.

The inhomogeneity of the magnetic field has a significant effect on the trajectories of charged particles in a tokamak. As they move along a helical magnetic field line, charged particles pass from the outer part of a toroidal magnetic surface to the inner, i.e., from a region with a weak magnetic field to one with a strong field, and back. During this motion the ratio of the parallel and perpendicular components of the velocity changes. Particles with small parallel velocities are trapped in the weak field region and oscillate between the mirror plugs formed by the field. The fraction of trapped particles is determined by the inhomogeneity in the field and is of order $\sqrt{2r/R}$. The trajectories of both the trapped particles and the particles which pass around the torus (known as passing particles) are displaced with respect to the magnetic surfaces because of toroidal drift. For passing particles, the "toroidal" projection of their trajectories on a transverse cross section of the torus is a circle whose center is displaced relative to the center of the magnetic surface. For the trapped particles, the toroidal projection of the drift trajectory has a closed sickle-like shape. (These trajectories are known as "banana" orbits.) The maximum displacement of this trajectory relative to the magnetic surface is considerably greater than the Larmor radius of the particles.

Particles with a small ratio of longitudinal to transverse velocity can also be trapped between the mirrors formed by ripples in the magnetic field. Such particles are referred to as "locally trapped." Toroidal drift of these particles leads to a larger displacement of their trajectories and, therefore, to greater losses.

The equilibrium conditions in a tokamak are related to the toroidal nonuniformity of the plasma column and magnetic field. This nonuniformity leads to the appearance of a force in the direction of the major radius of the torus. It is determined by a difference in the action of the kinetic and magnetic pressures on the inner and outer surfaces of the torus (ballooning effect) and by the electrodynamical repulsion

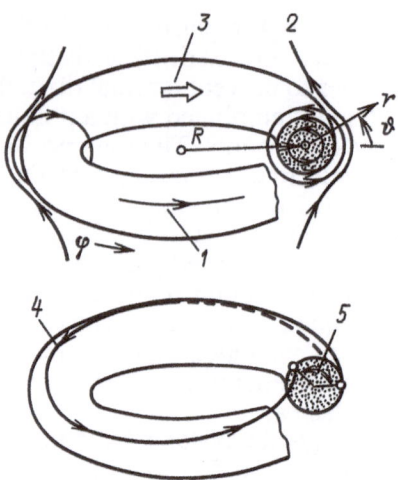

Fig. 1.3. The magnetic field configuration in a tokamak:
1) toroidal field; 2) poloidal field; 3) plasma current; 4) a magnetic field line; 5) the rotational transform angle.

of the toroidal current. A compensating force is created in a tokamak by the interaction of the vertical magnetic field B_v with the current. The equilibrium strength of
this field is of order $B_v \sim (a/R)B_\vartheta(a)$, where a and R are the minor and major radii
of the plasma column.

The equilibrium conditions also make it possible to determine the structure of
the magnetic surfaces in the plasma column. For very low aspect ratios a/R, the
structure of the magnetic surfaces corresponds to concentric circles in the merid-
ional cross section. As this ratio is increased while $a/R \ll 1$, the cross sections of
the magnetic surfaces are still close to circular, but their centers are displaced in the
direction of the major radius. This displacement is caused by an increase in the
distending force associated with the plasma pressure on moving from the peripheral
to the central magnetic surfaces. The shift of the magnetic axis from the centers of
the magnetic surfaces increases with the plasma pressure or, more precisely, with
the ratio of the average pressure in the plasma column to the poloidal magnetic field
pressure $\beta_p = 8\pi \overline{nT}/B_\vartheta^2$. Analysis shows that when $\beta_p \sim R/a$, the magnetic surfaces
are deformed. Their cross sections lose their circular shape and are "squeezed"
toward the outer edge of the torus.

We now briefly consider the conditions for stability of the equilibrium
configuration of the plasma in a tokamak. An analysis based on the equations of
magnetohydrodynamics (MHD) shows that instabilities caused by two types of de-
formation (helical and flute) can develop in an ideal plasma (with infinite
conductivity). Helical instabilities result from an excess of energy in the poloidal
magnetic field. Their maximum growth rate is determined by the Alfvén velocity v_A
and is usually extremely large compared to the reciprocal particle and energy life-
times of the plasma. The different helical perturbation modes can be characterized by
the numbers m and n, which determine the variation of the fields in the poloidal and

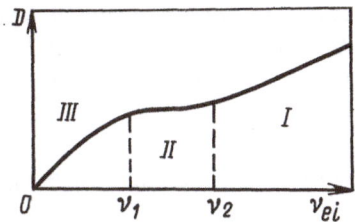

Fig. 1.4. The dependence of the diffusion coefficient of a tokamak plasma on the electron–ion collision frequency: $\nu_1 = (r/R)^{3/2}\nu_2$; $\nu_2 = (\nu_T/r)B_\vartheta/B\varphi$; I) hydrodynamic region; II) plateau; (III) banana region.

toroidal directions, $\xi = \xi_0 \exp[i(m\vartheta + n\varphi)]$. For helical instabilities to develop in an ideal plasma it is necessary that there be a surface in the vacuum region outside the plasma column on which the step size for a magnetic field line is equal to that for a helical perturbation. This condition can be expressed in terms of the safety factor (q) of the plasma as

$$q(r) = B_\varphi r/ (B_\vartheta(r)R).$$

Therefore, perturbations of a given mode are stable if $q(a) > m/n$ (the Kruskal–Shafranov criterion) near the plasma boundary. Depending on the current density distribution, perturbations with $m > 2$ may also be stable for $1 < q(a) < m$. For smooth current distributions, stability "windows" appear between rational values of $q(a)$. Taking the finite resistivity of the plasma into account reduces the limitations associated with "freezing" of the magnetic field in the plasma. This is especially important near the "rational" magnetic surfaces $q(r) = m/n$, where breaking and reconnection of magnetic field lines can occur and magnetic islands may form (the so-called "tearing mode"). Field line reconnection is especially important for the development of $m = 1$ instabilities in the neighborhood of $q(r_s) = 1$. Then the reconnection encompasses the entire core of the plasma, in which $q < 1$ until q becomes greater than unity throughout the region with $r < r_s$. As a result, efficient mixing of the plasma occurs there.

The second type of perturbations which may be unstable are the so-called "flute instabilities." They are caused by the plasma pressure gradient and develop inside the plasma column near the resonant magnetic surfaces $q(r) = m/n$. In a region with low shear ($r\, dq/dr < 1$), flute instabilities are stabilized by a magnetic well with $q > 1$. On the edge with strong shear, however, ballooning modes can develop on the low magnetic field side. These modes limit the maximum plasma pressure to the level of $\bar{\beta}_p < (0.3–0.5)R/a$. Including finite resistivity in the calculation may make the condition for stability relative to ballooning modes even more restrictive.

Other instabilities besides these MHD instabilities may affect the plasma behavior, in particular drift and kinetic instabilities. We shall not discuss them here, however.

In a stable plasma, charged particles and their energy are transported across the magnetic field by means of collisions. In a uniform magnetic field the collisions lead to displacements of the Larmor centers of the charged particles on the order of the Larmor radius. Thus, the diffusion coefficient D and thermal diffusivity χ, which characterize transport across a uniform field (known as "classical transport"), have the following magnitudes:

$$D_e \simeq D_i \simeq \chi_e \simeq \nu_{ei} \overline{\rho_{Be}^2}, \quad \chi_i \simeq \nu_{ii} \overline{\rho_{Bi}^2},$$

where ν_{ei} and ν_{ii} are the effective electron–ion and ion–ion collision frequencies and ρ_{Be} and ρ_{Bi} are the electron and ion Larmor radii.

In the nonuniform magnetic field of a tokamak, the collisional transport behavior varies significantly. (In this case the transport is referred to as "neoclassical.") At collision frequencies much greater than the bounce frequency of the particles on their drift trajectories, these trajectories are distorted and the displacement during collisions is of the same order as in a uniform field. The losses owing to toroidal drift of the particles in the residual electric field produced by charge separation, however, become significant since the shunting of the field caused by the longitudinal conductivity of the plasma is reduced as the collision frequency increases. Hence, the overall transport coefficients are somewhat larger than in a uniform field (by roughly a factor of q^2). At low collision frequencies, below the bounce frequencies of trapped and passing particles, transport occurs because a collision leads to "jumping" of particles from one drift trajectory to an adjacent one. (This is the so-called "banana" regime.) The displacement during such a jump is on the order of the width of a drift trajectory, which is much greater than the Larmor radius. Thus, the transport coefficients are also considerably higher than in a uniform field (roughly by a factor of $q^2 (R/r)^{3/2}$). The overall dependence of the transport coefficients on the collision frequency, shown schematically in Fig. 1.4, includes these two limiting cases and a plateau transition region.

At thermonuclear densities and temperatures, the effective collision frequency is low and neoclassical transport corresponds to the "banana" regime. The energy confinement time determined by this transport, $\tau_E \sim a^2/\chi_i$, is greater than that required for the Lawson criterion (1.3) when $I > 3(a/RT)^{1/4}$ (where T is in keV), i.e., for currents $I > 1$–1.5 MA. Meeting this condition, however, is not sufficient to ensure the required plasma confinement, since collisional (neoclassical) transport is supplemented by additional (anomalous) transport of particles and energy owing to oscillations in the plasma. A whole series of theoretical models has been proposed for anomalous transport. (See [12–14] and the literature cited there.) Up to now, however, the agreement between these models and experiment cannot be regarded as definitive.

1.3. EXPERIMENTAL RESEARCH ON TOKAMAKS

Experiments on tokamaks were begun at the Kurchatov Institute of Atomic Energy during the mid 1950s. In the course of tokamak development, technical difficulties were overcome and the concepts of plasma behavior in tokamaks were gradually formulated. As a result, by the end of the 1960s it was possible to obtain a tokamak plasma with a fairly high temperature and density and a relatively long en-

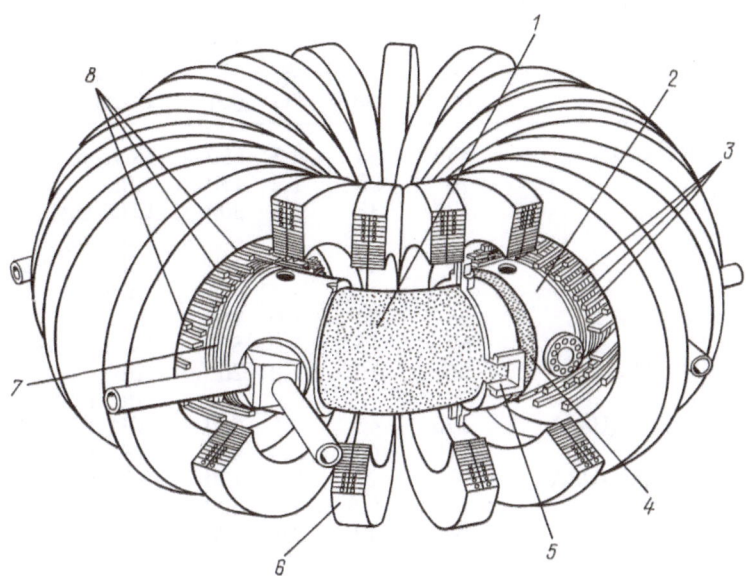

Fig. 1.5. A sketch of the PLT tokamak [13]: 1) plasma; 2) vacuum vessel; 3) vertical field coil; 4) ceramic break; 5) pumping duct; 6) toroidal field coils; 7) bellows section; 8) primary transformer winding.

ergy confinement time. Subsequently, research on tokamaks was undertaken in many laboratories around the world. Their successful development has meant that tokamaks occupy the leading position in research on controlled thermonuclear fusion [11–14]. Over 100 experimental tokamaks are operating in different laboratories. The parameters of the experiments in which auxiliary heating has been studied are listed in Table 1.1.

The layout of a typical tokamak experimental system is shown in Fig. 1.5.

The toroidal magnetic field in a tokamak is usually created by a multiturn solenoid. A toroidal current is induced in the plasma by a transformer in which the plasma column is the secondary winding. In most tokamaks an iron core transformer is used, although some modern systems use an air core transformer to obtain higher currents. To ensure equilibrium of the plasma column, special windings are installed in the tokamak to create vertical and horizontal magnetic fields. A thick metal shell which maintained a stabilizing field through Foucault currents was used to support the equilibrium in the first tokamaks. A shell is still used in some modern machines. In many tokamaks, the equilibrium field is maintained by a feedback system in which the magnetic probes used to determine the position of the plasma column serve as sensors.

A tokamak vacuum vessel is usually made of stainless steel or alloys with a similar composition. One or more limiters which establish the cross section of the plasma column are placed in the vessel. Limiters are made of various materials, including stainless steel, molybdenum, and graphite. The entry of impurities into the plasma is determined, to a large degree, by the interaction between the plasma and

Table 1.1. Parameters of Tokamaks

Tokamak	Laboratory, Country	R (cm)	a (cm)	B (T)	I (kA)	Auxiliary heating, Features
T-10	IAE (USSR)	150	36	4.5	650	ECH, ICH
T-11	"	70	20	1.2	150	NBI
T-7	"	122	35	2.5	300	LHH, superconducting coils
TO-2	"	60	14	2	40	ICH, toroidal diverter
Tuman-3	PTI (USSR)	55	24	1.0	150	AC, ICH
FT-1	"	62	15	1.0	50	ECH, LHH
R-05	SPTI (USSR)	65	10	1.0	20	Alfvén heating, quartz vacuum vessel
PLT	Princeton (USA)	130	40	3.5	550	NBI, ICH, LHH
PDX	"	140	40	2.4	450	NBI, ECH, PD
ISX-B	Oak Ridge (USA)	93	27	2	120	NBI, ECH
Alcator A	MIT (USA)	54	9.5	9	300	LHH
Alcator C	"	64	17	10	700	LHH, ICH
Versator II	"	40	15	0.8	40	LHH, ECH
Doublet III	General Atomic San Diego (USA)	140	45	2.6	1000	NBI, ECH, non-circular cross section, PD
TFR-600	Fontenay-aux-Roses (France)	98	20	5	400	NBI, ICH
Petula B	Grenoble (France)	72	15	2.7	80	LHH
Wega	"	72	15	2.5	50	LHH
DITE	Culham (UK)	112	23	2.8	200	NBI, diverter
Tosca	"	30	8.5	0.6	20	ECH, Alfvén heating
CLEO	"	90	18	1.5	140	ECH
JFT-2	JAERI (Japan)	90	25	1.8	150	LHH, ECH, ICH
JIPP-T II	Nagoya (Japan)	91	17	3.0	160	LHH, ICH, ECH
WT-2	Kyoto (Japan)	40	9	1.5	30	LHH, ECH
Asdex	Garching (FRG)	165	40	2.5	400	NBI, LHH, ICH, PD
FT	Frascati (Italy)	83	18	8	600	LHH
TCA	Lausanne (Switzerland)	61	18	1.5	140	Alfvén heating

Notes: IAE (I.V. Kurchatov Inst. of Atomic Energy); PTI (A.F. Ioffe Physicotechnical Inst.); SPTI (Sukhumi Physico Technical Inst.); MIT (Massachusetts Inst. of Technology); ECH (electron cyclotron heating); ICH (ion cyclotron heating); NBI (neutral beam injection); LHH (lower hybrid heating); AC (adiabatic compression); PD (poloidal diverter).

the limiter. The working gas (deuterium, hydrogen) is usually fed into the vessel through a pulsed valve which makes it possible to control the influx of gas during the discharge. An important part of tokamak technique is the cleaning and outgassing of the vessel walls. The methods used for this purpose include heating the vessel walls, titanium gettering, and prolonged cleaning by inductive and glow discharges in inert gases, oxygen, and hydrogen.

Table 1.2. Basic Plasma Diagnostics for Tokamaks

Measured quantity	Diagnostic types	Features
Electron density, n	Microwave interferometry IR interferometry	$\bar{n}$ (along chords) "
	Thomson scattering	$n(r)$ at one time

Electron temperature,
  T_e
	Thomson scattering
	Soft x-ray spectrum
		$T_e(r)$ at one time
		$T_e(t)$, electron
		  distribution function

	Second harmonic electron cyclotron emission	$T_e(r, t)$

Ion temperature, T_i	Energy spectrum of
	  charge exchange atoms
		$T_i(t)$, ion distribution
		  function

	Doppler broadening of impurity lines	$T_i(r)$
	Neutron flux	T_i for a maxwellian distribution
Neutral atom density, n_a	Charge exchange flux Resonance fluorescence	Averaged n_a $n_a(r)$
Impurity atom density, n_I	Soft x-ray and VUV spectra	n_I
Radiative power loss, W	Bolometry	$W(r)$

Stored energy and
  plasma pressure
	Diamagnetic measurements
	Measurements of equilib-
	  rium conditions
		–
		–

Spectrum of plasma oscillations	Magnetic probes Microwave scattering IR scattering	Edge measurements Localized measurements " "

Edge plasma param-
  eters
	Electrostatic probes
	Spectroscopy
	Resonance fluorescence
	  of hydrogen and impurities
	Microwave interferometry
	  and reflectometry
		Limiter "shadow"
		$\bar{n}_I, \bar{n}_a$
		$n_I(r), n_a(r)$
		
		$n(r)$

A large selection of diagnostic devices, some of which were specially developed for the purpose [15, 16], are used in tokamak experiments (Table 1.2).

Magnetic probes and loops are extensively used on tokamaks to determine the location of the plasma column, measure the diamagnetic effect proportional to the plasma pressure, and study the MHD oscillations of the plasma. Interferometric methods, based on the density dependence of the refractive index of the plasma, are used in the millimeter, submillimeter, and infrared ranges to determine the plasma density. Chordal measurements by these methods yield information on the line integral of the density along the path. Radial density profiles are derived from them by Abel inversion. Local information on the electron temperature is obtained by measuring the spectrum of Thomson scattered laser light. The temperature in the scattering volume is determined from the broadening of the scattered spectrum, and the electron density, from the scattering amplitude. Scattering measurements give the electron temperature and density at a single time. In order to study the evolution of the electron temperature, measurements are made of the spectra of the plasma emission near the second harmonic of the electron cyclotron frequency and of the soft x-ray emission. The slope of the x-ray spectrum (owing to bremsstrahlung and recombination) yields the electron temperature in the hottest region along the line of sight. The emission near the second harmonic of the cyclotron frequency is usually "black" and is determined by the electron temperature within a narrow plasma region at the intersection of the $\omega = 2\omega_{Be}$ surface and the directional diagram of the receiver antenna. The radiation temperature measurements are often calibrated to the Thomson scattering measurements at a single time. Information on the ion temperature is usually obtained by measuring the energy spectrum of the fast neutral atoms formed by charge exchange of ions in the plasma volume. Yet another method of determining the ion temperature is based on measuring the Doppler broadening of spectrum lines of impurity ions. The resulting data are localized because ions with a given charge are usually concentrated in a fairly narrow plasma region set by the conditions under which they are ionized. In some cases, the ion temperature is found from measurements of the intensity of (D, D) neutron reactions. Information about impurities is obtained by measuring the characteristic plasma emission lines in the vacuum ultraviolet and soft x-ray ranges. The total radiated energy loss is measured by bolometers. The effective charge Z_{eff} of the plasma is related to the density and charge of impurity ions and is found from the longitudinal conductivity of the plasma. The set of measurements of the electron and ion temperatures and densities, together with the radiative energy loss, makes it possible to construct equations for particle and energy balance. A detailed analysis of these equations is made by numerical modelling using so-called transport codes.

We now briefly consider the results of experiments on ohmically heated tokamak plasmas. Equilibrium and MHD stable discharge conditions can be obtained over a limited range of discharge currents and plasma densities. The lowest density (usually $n \sim 10^{-3}$ cm^{-3}) is determined by a transition to an unstable regime with a large number of runaway electrons. The highest density corresponds to a transition into an unstable regime in which the discharge current "disrupts." It is roughly given by the scaling law [17]

$$\bar{n}_{max} \approx (3-8) \cdot 10^{15} \, B/R, \qquad (1.7)$$

where $\bar{n}_{max}$ is in cm^{-3}, B is in T, and R is in cm. The nature of this transition has not yet been established.

The limiting plasma density depends on the gas feed regime. It increases when high auxiliary heating powers are applied to the plasma. The smallest value of q for stable operation, which corresponds to the highest current, depends on the cleanliness of the vessel. The influx of impurities is associated with a cooling of the plasma edge and, consequently, with a narrowing of the electron temperature and current profiles. The widest profiles and lowest values of q are obtained in very clean vessels. In many modern experiments, q_{min} values of 2–2.5 are obtained, and in some it has been possible to obtain still smaller values of 1.2–1.5. It is true that for $q < 2$ a considerable deterioration in the plasma energy confinement was noted.

The behavior of plasmas with a vertically elongated cross section has also been studied. This sort of D-shaped cross section is obtained with the aid of a special magnetic system. On going from a circular to a D-shaped cross section, higher values of the poloidal field and current, and, therefore, higher plasma pressures are obtained for fixed q because of the increased perimeter of the plasma column.

MHD oscillations are observed in different tokamak operating regimes by means of magnetic probes. As the current rises, oscillations with successively lower azimuthal wave numbers m, beginning with $m = 6$–8, are excited. In the quasistationary stage, modes with $m = 2$ and $m = 3$, probably corresponding to the tearing mode, are usually observed. The development of quasistationary MHD modes under certain conditions leads to disruption of the discharge current. This usually begins with the $m = 2$ mode and then, as the current decreases, higher modes are excited. In low q (less than 3–4) regimes, sawtooth relaxation oscillations in the temperature and density (the so-called internal disruption instability) are seen. These oscillations are observed inside the volume at whose minor radius $q = 1$. They correspond to a mode with $m = 1$ and $n = 1$. The development of the instability is accompanied by rapid mixing of the plasma, probably caused by reattachment of magnetic field lines.

Studies of the plasma energy and particle balance have received considerable attention in tokamak experiments. The energy balance is customarily characterized by the energy confinement time τ_E, which is defined in the steady state as the ratio of the energy stored in the plasma, $(3/2)\overline{n(T_e + T_i)}V$, to the heating power P_H:

$$\tau_E = (3/2)\,\overline{n\,(T_e + T_i)}\,V/P_H, \qquad (1.8)$$

where V is the plasma volume. The stored energy is averaged over this volume. The energy confinement times for the electrons and ions, τ_{Ee} and τ_{Ei}, are also distinguished. Data on the ion energy confinement time are in good agreement with the natural assumption that the losses are caused by removal of ions from the plasma volume through resonance charge exchange and by heat transport associated with ion thermal conduction. The ion thermal conductivity determined from the energy confinement time is of the same order of magnitude as the neoclassical value for the plateau region corresponding to a collision frequency greater than the bounce frequency of the trapped particles along a "banana" trajectory. The energy balance established by collisional energy transfer from the electrons to the ions and by energy losses corresponding to neoclassical ion heat conduction in the "banana" regime yields Artsimovich's [11] approximation for the ion temperature:

$$T_i \approx 1.3 \cdot 10^{-7} \sqrt[3]{IB_\varphi R^2 n_e}/\sqrt{A_i}, \qquad (1.9)$$

where T_i is in keV, I in kA, B in T, R in cm, and n_e in cm^{-3}. It is usually satisfied to within a factor of 2–3 for ohmic heating. A later, more detailed analysis showed that the experimental ion thermal conductivity is 2–8 times the neoclassical value. This leads to a reduction in T_i by a factor of 1.5–2 compared to the predictions of the neoclassical theory.

The electrons lose energy through radiation, primarily associated with impurities, and heat transfer. In "clean" tokamaks, the radiative losses over the entire volume are usually about 20–30% of the power applied to heat the plasma. The energy losses are mainly associated with heat transfer. The experimentally observed heat transfer greatly exceeds the neoclassical electron heat conduction. This heat transfer is referred to as "anomalous" and is related to plasma fluctuations. There is a theory that this heat transfer is caused by the breakup and reconnection of magnetic field lines over distances on the order of the collisionless skin depth, but no reliable interpretation of the anomalous transport has yet been found. Thus, considerable attention is devoted to the search for empirical relationships between the electron energy confinement time and the plasma parameters. Substantially different dependences have been obtained in various experiments. It has been shown for a wide range of parameters in many ohmically heated tokamaks that the energy confinement time is proportional to the electron density. This dependence corresponds to the so-called Alcator scaling, given by (τ_E in s)

$$\tau_E \simeq (3 - 5) \cdot 10^{-19} \, \bar{n} a^2. \tag{1.10}$$

This leads to an electron thermal conductivity of $\kappa_e \simeq 3\text{–}5 \cdot 10^{17} \, \text{cm}^{-1} \cdot \text{s}^{-1}$. In recent years the dependence of τ_E on the minor radius a given by Eq. (1.10) has come into question. The dependence on a is apparently weaker while a strong dependence on the major radius R is observed. A large amount of data corresponds to the scaling observed on the T-11 and Alcator C tokamaks [18, 19]. The so-called neoalcator scaling is

$$\tau_E \simeq 10^{-21} \, \bar{n} a R^2 \sqrt{q}. \tag{1.11}$$

This corresponds to a dependence of the thermal conductivity on the machine geometry of the form $\kappa_e \sim r/R^2$.

It should be noted that at low q, the heat transfer in the central part of the plasma can be much faster than that given by Eqs. (1.10) and (1.11). Enhanced transport may be caused by rapid mixing of the plasma when magnetic field lines reconnect in a region with sawtooth oscillations, where $q \lesssim 1$. Faster transport is sometimes observed on the edge [with $q(r) > 2\text{–}3$]. Hence, the behavior described above characterizes heat transfer in the middle of the plasma, where most of the electron temperature drop occurs.

The relationship between the energy confinement time and the electron temperature cannot be determined in ohmic heating experiments because the heating power depends strongly on T_e. Judging from experiments with nonohmic auxiliary heating, the dependence of τ_E on T_e is not strong as long as β is low ($\bar{\beta}_p < 0.5$). Experiments with high auxiliary heating powers, on the other hand, indicate that the confinement time decreases at high β. Goldston [19] has proposed an energy confinement time scaling that includes a strong dependence on β: $\tau_E \sim I^2/\langle nT_e \rangle$. This scaling, however, can hardly be regarded as universal. Two confinement regimes (L and H modes) have been shown to exist in many experiments with high-power auxiliary heating [20]. The L mode is characterized by confinement times several times smaller than with ohmic heating. In the H mode the confinement time is close to that obtained with ohmic heating. This mode is achieved by reducing the influx

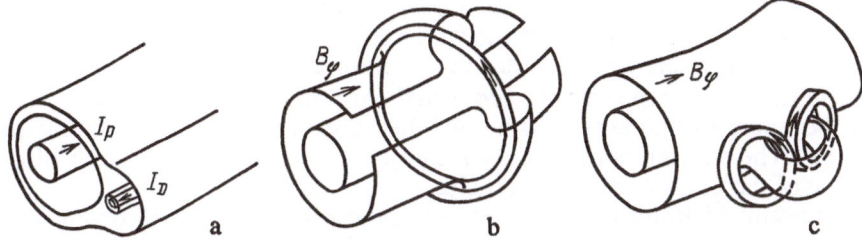

Fig. 1.6. Divertor schemes in tokamaks: a) poloidal; b) toroidal; c) bundle divertor. (I_D denotes the current in the divertor coil.)

Table 1.3. Plasma Parameters Obtained in Large Tokamaks

Tokamak	a (cm)	B (T)	I (kA)	Auxiliary heating	$\overline{n}$ (10^{13} cm^{-3})	$T_e(0)$ (keV)	$T_i(0)$ (keV)	τ_E (s)
T-10	30	3	300	ECH, P=1 MW	3	4	1	0.05
PLT	40	3.5	400	NBI, P=3 MW	3	3	5	0.04
				ICH, P=3 MW	3	3	3	0.04
Alcator C	17	10	600		10	1.5	1.5	0.05
Doublet III	45	2.5	800	NBI, P=8 MW	7	4	5	0.08
Asdex	40	2.5	400	NBI, P=4 MW	2	3	3	0.06
FT	18	8	140	–	10	1.5	1.4	0.05
TFTR	85	3	1400	–	3	2	2	0.3
JET	125	3.4	3200	–	3	3	2	0.8

Table 1.4. Design Parameters of the Latest Tokamaks

Tokamak	Laboratory, country	R (cm)	a (cm)	B (T)	I (MA)	τ_I (s)	Auxiliary heating	Features
TFTR	Princeton (USA)	248	85	5	2.5	1.5	NBI, ICH AC	–
JET	Joint European Torus (UK)	296	125× 210*	3.4	4.8	15	NBI, ICH	D-shaped cross section
JT-60	JAERI (Japan)	300	100	5	3.0	10	NBI, LHH	Diverter
T-15	IAE (USSR)	240	70	5	2.0	5	NBI, ECH	Superconducting coils
Torus II	Cadarache (France)	215	75	4.5	1.7	30	NBI, ICH	The same
"Intor"	International project project	520	120× 200*	5.5	6.4	200.	NBI	D-shaped cross section, Superconducting coils

* Half width and half height of D-shaped cross section.

of impurities into the plasma and forming a flatter radial electron temperature profile. The nature of transport in these confinement regimes is still unclear.

The transport of charged particles in the plasma is usually characterized by the particle lifetime τ_p. Experiments show that this time is greater by a factor of 2–10 than the energy confinement time and has a similar dependence on the plasma parameters. The confinement time for impurity ions is of a similar order of magnitude to that of the primary plasma particles.

The amount of impurity atoms entering the plasma depends strongly on the outgassing and cleaning of the vessel walls. Modern conditioning techniques make it possible to reach comparatively low values of the effective charge, defined as

$$Z_{eff} = \sum_j n_j Z_j^2 / \sum_j n_j Z_j, \qquad (1.12)$$

where Z_j is the ionic charge of the jth impurity, n_j is the ion density of this impurity, and the sum is taken over all ionization states and all types of ions. In the cleanest tokamaks $Z_{eff} = 1$–1.5. In a thermonuclear reactor the rate of influx of high-Z impurities must be low enough for radiation losses to be small compared to the thermonuclear yield. In this regard, methods based on the use of divertors are being developed to reduce the interaction of the plasma with the wall. A divertor configuration for the magnetic field (Fig. 1.6) ensures that the edge plasma "drains off" into a special chamber where the plasma comes into contact with the wall. This makes it possible to greatly reduce the plasma temperature in the contact region and efficiently pump out the gas released from the wall.

We conclude with a list of the plasma parameters obtained in several large tokamak experiments through the end of 1984 (Table 1.3).

Experiments are now beginning on the next generation of tokamaks in which ignition (physical proof of principle) is expected to occur. Two of these tokamaks, TFTR and JET, began operation in 1983. The plasma parameters obtained in them are given in Table 1.3 [21, 22]. Other machines will probably begin operation in the next 2–3 years. The next step after these tokamaks should involve the construction of an experimental thermonuclear reactor with basic engineering systems. An international design for such a reactor "Intor" has been prepared [23] (see Table 1.4).

1.4. STELLARATORS

Stellarators, like tokamaks, are toroidal magnetic confinement systems [24, 25]. In stellarators, as opposed to tokamaks, however, the poloidal component of the magnetic field, as well as the toroidal, is created by currents in external magnetic field windings. Hence, toroidal magnetic surfaces exist in stellarators even in the absence of plasma. The first stellarators were built in Princeton (USA) in the 1950s. The results obtained on them were not promising, for the plasma confinement characteristics were considerably worse than in tokamaks. Interest in stel-

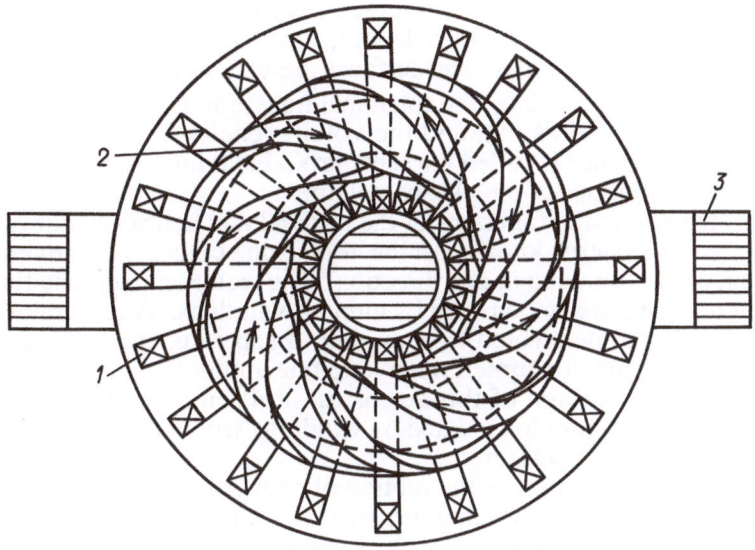

Fig. 1.7. A sketch of an $l = 3$ stellarator [24]: 1) toroidal field coil; 2) helical windings; 3) iron core for current excitation.

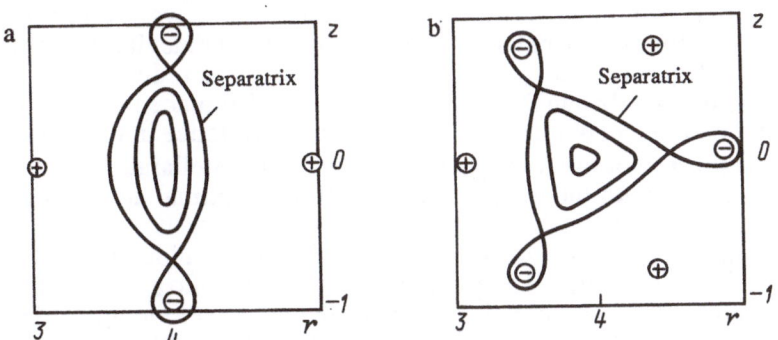

Fig. 1.8. Transverse cross sections of the magnetic surfaces in $l = 2$ (a) and $l = 3$ (b) stellarators [24]. The helical windings are shown by circles with + and − signs.

larators has developed anew in recent years after successful studies in the USSR, West Germany, and Japan. Although the scale of the experiments is considerably greater on tokamaks than on stellarators, the latter are now viewed as a possible alternative.

In most experiments until now the stellarator magnetic field configuration has been created by a combination of a toroidal solenoid and a separate helical winding (Fig. 1.7). The helical winding consists of $2l$ conductors where the current in each pair of conductors flows in opposite directions. Usually double or triple turn windings ($l = 2$ or 3) are used. The poloidal field created by these windings, together with the toroidal field of the solenoid, forms magnetic surfaces inside the stellarator (Fig. 1.8). These magnetic surfaces are characterized by the step size L of the helical magnetic field lines, the rotational transform angle $\iota = \Pi/L$ (where Π is the length of the magnetic axis), and the volume enclosed by the surface. The boundary magnetic surface is called the separatrix. Field lines outside the separatrix extend beyond the confines of the plasma volume.

In some stellarators the functions of the toroidal and helical windings are combined. One such system, known as the torsatron, consists of several helical windings with current flowing in one direction. The resulting magnetic flux parallel to the axis of the torus is compensated by circular windings. A stellarator magnetic configuration can also be created by a system of separate, specially shaped coils (for example, elliptical coils turned relative to one another). From an engineering standpoint, a modular design of this sort is most efficient for large devices. Finally, a stellarator configuration created by windings with a three-dimensional (not plane) axis is under consideration.

A theoretical examination of the motion of charged plasma particles in a stellarator field configuration shows that, because of the lack of axial symmetry, their trajectories are more complicated than in a tokamak. The "banana" orbits themselves move in the toroidal cross section, forming so-called "superbananas" associated with particles trapped between the maxima of the helical field. As a result, neoclassical transport at low collision frequencies (below the "bounce" frequency of the particles along the "bananas") can be considerably greater than transport in tokamaks. Enhanced transport makes attainment of the Lawson criterion more difficult. Incidentally, numerical simulations show that the deviations from an ideal stellarator configuration in real systems may cause a reduction in the effects associated with "superbanana" transport [26]. The difference in the MHD stability of stellarator and tokamak plasmas is caused by a difference in the radial profile of the safety factor q or of the rotational transform ι. In a stellarator without a plasma current, q is greatest on the magnetic axis. With an optimal design for the helical windings, the principal instabilities can be suppressed.

The limiting plasma pressure in stellarators is determined by the equilibrium conditions associated with the displacement of the magnetic axis or with distortions in the magnetic surfaces and by the stability relative to ballooning modes. According to theoretical predictions, values of $\beta = 8\pi\overline{nT}/B^2$ higher than those in tokamaks can be obtained with an optimum aspect ratio. The highest values of β (up to 20%) can apparently be reached in stellarators with a three-dimensional axis.

Until recently, experiments in stellarators were conducted with plasmas that were created and heated by a toroidal current. Under these conditions the differences between modern stellarators and tokamaks did not seem very great. The range of currents for which MHD stable operation is obtained is determined by the Kruskal–Shafranov criterion, which takes into account the total poloidal field produced by the external helical windings and the discharge current. In stellarators, as in tokamaks, helical MHD modes and disruptive instabilities are observed. The ion energy confinement is close to neoclassical. Electron heat transfer and particle transport are anomalous, with behavior that is close to that observed in tokamaks. All these results have been obtained at low β ($\overline{\beta} < 0.2\%$).

Table 1.5. Parameters of Stellarators

Machine	Location	R (cm)	a (cm)	B (T)	l/N	max	Auxiliary heating
L-2	IGP (USSR)	100	11.5	2	2/14	0.78	ICH, ECH
Uragan-2	KPTI (USSR)	110	6.3	2	3/18	1	ICH
Uragan-3	"	100	7	1	3/3	0.7	ICH
W-VIIA	Garching (FRG)	200	10	3.5	2/5	0.23	ICH, ECH
Heliotron E (rectangular cross section)	Kyoto (Japan)	220	15× 30	2	2/19	2.5	NBI, ECH, ICH

Notes: IGP (Institute of General Physics, USSR Academy of Sciences); KPTI (Kharkov Physico-technical Institute).

Stellarator experiments without a toroidal current are comparatively recent. In these experiments, the plasma is created and maintained by neutral beam injection and electron cyclotron heating. Peak values of $\bar{\beta} \approx 2\%$ have been obtained. The initial experiments indicate that electron thermal transport is greatly reduced in the currentless regime.

The parameters of several modern stellarators are listed in Table 1.5. The construction of the next generation of larger experiments is now under way, and reactor designs based on the stellarator are being considered. The main advantages of such a reactor over the tokamak result from the absence of a toroidal current, which may permit steady-state operation.

1.5. PLASMA HEATING EFFICIENCY

Heating in the present and next generation of experiments can be evaluated in terms of experimental relationships for the energy confinement time τ_E. The plasma electron temperature in an ohmically heated tokamak can be obtained from the steady-state energy balance condition:

$$P_{OH} = (3/2)\,\overline{nT_e}\,V/\tau_{Ee}. \tag{1.13}$$

The ohmic heating power P_{OH} released in the plasma is determined by the electrical resistance $\mathcal{R}$ of the toroidal plasma column:

$$P_{OH} = I^2 \mathcal{R} = I^2 2\pi R/(\pi a^2 \bar{\sigma}), \tag{1.14}$$

where the specific resistivity averaged over the cross section of the plasma, $\bar{\sigma}$, is given by

$$\bar{\sigma} = \frac{3}{2\sqrt{2\pi}} \frac{\overline{T_e^{3/2}}}{Z_{eff}e^2 m_e^{1/2} \Lambda_e \zeta}. \tag{1.15}$$

Here Z_{eff} is the effective ionic charge, Λ_e is the Coulomb logarithm, and ζ is the neoclassical correction to the conductivity. (For the parameters of modern experiments, $\Lambda_e = 10$–15 and $\zeta = 1.1$–1.5.) The peak current is limited by the Kruskal–Shafranov stability condition and can be expressed in terms of the safety factor q as

$$I = cB_\varphi a^2/(2qR). \tag{1.16}$$

Substituting this value into Eq. (1.14), we obtain

$$P_{OH} = \frac{4\sqrt{2\pi}}{3} \zeta a_\sigma \frac{m_e^{1/2} c^2 Z_{eff} e^2 a^2 B^2}{q^2 R T_{e\,0}^{3/2}} \tag{1.17}$$

or, numerically,

$$P_{OH} \simeq 1.15 \, \zeta a_\sigma Z_{eff} \frac{a^2}{R} \frac{B^2}{q^2 T_{e\,0}^{3/2}},$$

where $a_\sigma = T_{e0}^{3/2}/\overline{T_e^{3/2}}$ is a factor determined by the temperature averaging in Eq. (1.17), P_{OH} is in MW, a and R are in m, B is in T, and T_{e0} in keV.

Equation (1.17) for the ohmic heating power can now be put into the energy balance equation. Using the scaling law (1.11) for τ_E and assuming that the energy losses through radiation and electron–ion energy transfer are unimportant, Eq. (1.13) yields

$$T_e \approx 0.7 g a^{2/5} B^{4/5}/q^{3/5}, \tag{1.18}$$

where $g = (a_\sigma a_p Z_{eff}\zeta)^{2/5}$ is a numerical factor which usually lies between 1 and 1.5, and $a_p = n(0)T_e(0)/(\overline{nT_e})$. For modern tokamaks this formula gives $T_e = 0.5$–1.5 keV.

The efficiency of auxiliary nonohmic heating depends strongly on the energy losses associated with the heating mechanism. When the additional heating takes place through the electrons and the power transferred from the electrons to the ions is small, the steady-state energy balance condition can be approximated in a form analogous to Eq. (1.13):

$$P_{OH} + \gamma P_{AH} = (3/2)\,\overline{n\,(T_e + \Delta T_e)}\, V/\tau_{Ee}, \tag{1.19}$$

where P_{AH} is the auxiliary heating power, γ is the fraction of this power absorbed in the plasma, ΔT_e is the increase in the electron temperature during this heating, and V is the volume of the heated plasma. Assuming that τ_{Ee} scales with auxiliary heating according to the ohmic scaling (1.11), substituting it into Eq. (1.19) yields

$$\Delta T_e = 0.2\sqrt{q}\,(R/a)\,(\gamma P_{AH} + P'_{OH} - P_{OH}),\tag{1.20}$$

where P_{OH}' is the ohmic heating power during the auxiliary heating period, ΔT_e is in keV, and P_{AH} and P_{OH} are in MW.

In order to evaluate the heating efficiency in general, a more detailed analysis of the energy balance is needed. The energy balance equations for the electrons and ions for a given volume of plasma can be written in the form

$$\left.\begin{aligned}
\frac{dW_e}{dt} &= P_{OH} + \gamma_e P_{AH} - \frac{W_e}{\tau_{Ee}} - \frac{(W_e - W_i)}{\tau_{ei}}, \\[2mm]
\frac{dW_i}{dt} &= \gamma_i P_{AH} - \frac{W_i}{\tau_{Ei}} + \frac{(W_e - W_i)}{\tau_{ei}}.
\end{aligned}\right\}\tag{1.21}$$

The sum of these equations determines the energy balance of the plasma as a whole:

$$dW/dt = P_{OH} + \gamma P_{AH} - W/\tau_E.\tag{1.22}$$

In Eqs. (1.21) $W_e = (3/2)\bar{n}T_e V$ is the energy of the plasma electrons in volume V^*; $W_i = (3/2)\bar{n}_i\bar{T}_i V$ is the energy of the plasma ions; $W = W_e + W_i$; γ_e and γ_i are the fractions of the auxiliary heating power expended in heating the plasma electrons and ions in volume V; $\gamma = \gamma_e + \gamma_i$; τ_{Ee} is the average electron energy confinement time including transport losses associated with heat conduction and diffusion, as well as radiation; τ_{Ei} is the ion energy confinement time including transport and charge exchange losses; τ_E is the energy confinement time of the plasma as a whole; and τ_{ei} is the electron–ion collisional energy exchange time, given by

$$\tau_{ei} = \frac{3}{8\sqrt{2\pi}}\,\frac{m_i T_e^{3/2}}{m_e^{1/2} n e^4 \Lambda},\tag{1.23}$$

or, numerically, by

$$\tau_{ei} \simeq 6.7\cdot 10^{11}\,A_i T_e^{3/2}/n,$$

(with τ_{ei} in s, T_e in keV, and n in cm^{-3}).

In the steady state, Eqs. (1.21) yield the following expressions for W_e, W_i, and W:

$$\left.\begin{aligned}
W_e &= \frac{(P_{OH} + \gamma_e P_{AH})\tau_{Ee}\,(\tau_{Ei} + \tau_{ei}) + \gamma_i P_{AH}\tau_{Ee}\tau_{Ei}}{\tau_{Ee} + \tau_{Ei} + \tau_{ei}}, \\[2mm]
W_i &= \frac{(P_{OH} + \gamma_e P_{AH})\,\tau_{Ee}\tau_{Ei} + \gamma_i P_{AH}\tau_{Ei}\,(\tau_{Ee} + \tau_{ei})}{\tau_{Ee} + \tau_{Ei} + \tau_{ei}};
\end{aligned}\right\}\tag{1.24}$$

$^*V$ may be the entire plasma volume or the volume of the hot central part of the plasma column.

$$W = (P_{OH} + \gamma P_{AH}) \tau_E, \tag{1.25}$$

where τ_E is determined by a combination of the three times, τ_{Ee}, τ_{Ei}, and τ_{ei}, in the energy balance equation.

The experimental determination of the efficiency of auxiliary heating is an important question. It is often found by comparing the stored energy in the plasma during ohmic heating alone, W_{OH}, to the stored energy when the auxiliary heating is turned on. For the steady state, Eq. (1.25) gives

$$\frac{W}{W_{OH}} = \frac{\rho_h P_{AH} + P'_{OH}}{P_{OH}} = \frac{\gamma P_{AH} + P'_{OH}}{P_{OH}} \frac{\tau'_E}{\tau_E}, \tag{1.26}$$

where P_{OH} and τ_E refer to the ohmically heated plasma and P_{OH}' and τ_E', to combined ohmic and auxiliary heating. The changes in these quantities with auxiliary heating is related to changes in the plasma parameters, and τ_E' also depends on the distribution of heating power over the plasma cross section. As can be seen from Eq. (1.26), the coefficient ρ_h characterizes the overall efficiency of auxiliary heating compared to ohmic heating:

$$\rho_h = \gamma \frac{\tau'_E}{\tau_E} + \frac{P'_{OH}}{P_{AH}} \left(\frac{\tau'_E}{\tau_E} - 1 \right), \tag{1.27}$$

which, for $P_{OH} \ll P_{AH}$, gives

$$\rho_h = \gamma \tau'_E / \tau_E.$$

In many experimental studies a normalized heating parameter η_h, defined in terms of the ratio of the temperature rise in the plasma center to the auxiliary heating power, is introduced:

$$\eta_h = \bar{n} \Delta\, (T_{e0} + T_{i0}) / P_{AH}. \tag{1.28}$$

Using Eq. (1.25), we obtain

$$\eta_h \simeq 4.1 \cdot 10^{21} \left[\alpha'_p \frac{\gamma \tau'_E}{V} - \frac{(\alpha_p P_{OH} \tau_E - \alpha'_p P'_{OH} \tau'_E)}{P_{AH} V} \right], \tag{1.29}$$

where $\alpha_p = \bar{n}(T_{e0} + T_{i0}) / \overline{n(T_e + T_i)}$ and η_h is given in eV/(kW·cm³). When $P_{AH} \gg P_{OH}$, only the first term in Eq. (1.29) remains. Then

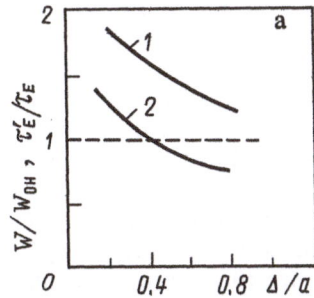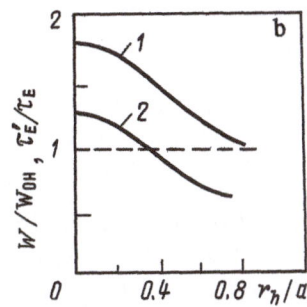

Fig. 1.9. The dependence of the energy content (1) and energy confinement time (2) of a plasma on the localization of auxiliary heating [28] ($P_{OH} = P_{AH}$): a) variation with the width Δ of the heating zone for $r_h = 0$; b) variation with the radial position r_h of the heating zone for $\Delta = 0.2a$.

$$\eta_h/\rho_h \simeq 4.1\cdot10^{21}\, \alpha'_p \tau_E/V, \qquad (1.30)$$

or with the scaling law (1.11) for τ_E,

$$\eta_h/\rho_h \simeq 0.2\, \alpha'_p \bar{n}(R/a)\sqrt{q}. \qquad (1.31)$$

Other definitions of the heating efficiency are possible. All of them are related to γ, the fraction of energy absorbed in a specified plasma volume, and to τ_E', the energy confinement time characteristic of the auxiliary heating. It should be noted that there is an independent experimental method for determining these times based on the time-dependent heating phases. During the initial phase of pulsed heating, the energy balance equation (1.25) can be written in the form

$$dW/dt = \gamma P_{AH} + P_{OH} - W/\tau_E \approx \gamma P_{AH}, \qquad (1.32)$$

since the plasma parameters have not yet changed significantly relative to the steady-state ohmic heating phase and $P_{OH} \sim W/\tau_E$. Then γ can be determined experimentally from the rate of rise of W [5]. τ_E' can be evaluated during the initial drop in W after the heating pulse is turned off, when the energy balance equation has the form

$$dW/dt \approx P'_{OH} - W/\tau'_E, \qquad (1.33)$$

and P_{OH}' and τ_E' are the same as during the auxiliary heating phase. Similarly, it is possible to find γ_e, γ_i, τ_{Ee}', and τ_{Ei}' from the rise and fall in W_e and W_i at the beginning and end of the heating pulse.

The approach considered here is based on the overall energy balance equations. It should be kept in mind that the confinement time for the heating energy, which determines the heating efficiency, depends strongly on the spatial distribution of the heat input. Heating is obviously most efficient in the plasma center. During edge heating a significant fraction of the energy applied to the plasma will escape to the walls because of heat conduction. The effect of the radial variation in the heat input on the heating efficiency can be analyzed by numerical simulation of transport processes using so-called transport codes. As an example, Fig. 1.9 shows the results of a heating calculation for a cylindrical plasma model with a thermal conductivity $\kappa(r) = \kappa_0(1 + 4r^2/a^2)$ which increases toward the edge. The heating efficiency decreases substantially on going from central to edge heating.

In conclusion, we note that the effect of heating on the energy distribution functions of the charged particles requires special attention. It is easy to evaluate the conditions for a small perturbation in the bulk of the distribution function. Since the electron (ion) energy distribution is established by electron–electron (ion–ion) collisions, this condition may be written in the form $\tau_{ee} \ll \tau_{Ee}$ for the electrons and $\tau_{ii} \ll \tau_{Ei}$ for the ions. (Here τ_{ee} and τ_{ii} are the averaged times for electron–electron and ion–ion collisional relaxation.) These inequalities are less restrictive than the conditions for sufficiently rapid energy exchange between the electrons and ions, since

$$\left.\begin{aligned}
\tau_{ee}/\tau_{ei} &\approx 2\sqrt{2}\, m_e/m_i, \\
\tau_{ii}/\tau_{ei} &\approx 2\sqrt{2}\sqrt{m_e/m_i}\, (T_i/T_e)^{3/2}.
\end{aligned}\right\} \tag{1.34}$$

They are usually satisfied in present-day experiments. The criteria for "Maxwellianization" of the energy distribution may be more restrictive if the heating energy is delivered to the "tail" of the distribution, since the collisional relaxation times increase with the energy E as $E^{3/2}$. Then the condition for efficient heating obviously requires that the time for energy transfer to the bulk plasma be considerably shorter than the confinement time for the group of heated particles.

1.6. SPECIFICATIONS FOR HEATING TECHNIQUES

We now consider the requirements for plasma heating in a toroidal (tokamak or stellarator) thermonuclear reactor. First, we estimate the power required to bring a reactor to ignition. Under quasistationary conditions it can be written in the form

$$P = \frac{3}{2}\, \frac{\overline{n\,(T_e + T_i)\, V}}{\tau_E} \approx \frac{\overline{3n^2 TV}}{n\tau_E}, \tag{1.35}$$

where n is the required density, T_e and T_i are the required temperatures, V is the effective plasma volume, and $\overline{n(T_e + T_i)} \approx \overline{2nT}$ is averaged over this volume. Using the ignition criterion (1.4) and the values of the density and temperature in a reactor (Section 1.1), we obtain

$$P \lesssim (0.15\text{--}0.3)V, \tag{1.36}$$

where P is in MW and V in m³. For proposed reactor designs, $P = 50\text{--}100$ MW. The duration of quasistationary heating evidently must exceed the energy confinement time. Using Eq. (1.4), we find that it must satisfy the condition

$$\tau > 3\cdot10^{14}/n\,(\text{cm}^{-3}) \approx 2\text{-}3 \text{ s}. \tag{1.37}$$

It is easy to confirm that ohmic heating by a toroidal current in practically attainable magnetic fields is not sufficient for reaching ignition since the current is limited by the stability condition and the plasma resistance drops with rising temperature. The ohmic heating power released in a plasma is given by Eq. (1.17). Substituting it in Eqs. (1.35) or (1.36), we obtain a criterion for ignition with ohmic heating:

$$B > (8-12) \, qR / \sqrt{\varsigma a_\sigma Z_{\text{eff}}}. \qquad (1.38)$$

This condition leads to a magnetic field ($B > 30$–50 T) which cannot be obtained in practice. For the fields in planned reactors ($B = 5$–8 T), the ohmic heating power ($P \sim B^2$) is 1.5–2 orders of magnitude lower than that required to reach ignition. Therefore, a tokamak requires additional nonohmic heating sources whose power must be much greater than the ohmic heating power. In a stellarator this heating may be done without having an ohmic heating current through the plasma.

The distribution of heating energy between the electrons and ions is determined by the time for collisional energy exchange between them (1.23). For the parameters of a thermonuclear reactor we have

$$n\tau_{ei} \approx 6.7 \cdot 10^{11} \, T_e^{3/2} A_i \approx 4 \cdot 10^{13} \ \text{cm}^{-3} \cdot \text{s}, \qquad (1.39)$$

where A_i is the atomic weight of the ions and T_e is in keV. Evidently, for thermonuclear conditions when the criterion (1.4) is satisfied, $\tau_{ei} < 0.2\tau_E$. Then the heating energy applied to the plasma should be almost equally distributed between the electrons and ions.

Several other requirements for plasma heating in thermonuclear systems should be mentioned. One important requirement pertains to localization of the region where heating power is delivered to the plasma. It is obviously best to heat the central region of the plasma. As noted above, when the heating is "distributed," especially at the edge, its efficiency is reduced.

One obvious requirement is that there not be a major increase in the influx of impurities from the walls during heating, which would cause increased radiation losses and other negative effects. For the same reason, the release of heating power in the edge plasma near the walls is also undesirable.

Yet another important requirement is that heating must not cause current disruptions, deterioration of plasma stability, or significant increases in the transport coefficients. Otherwise, higher heating powers will be necessary and the difficulties associated with impurity influx will become worse.

Finally, a necessary condition for the application of auxiliary heating to thermonuclear systems is that the heating equipment must be compatible with the vacuum vessel and magnetic system. In the case of a thermonuclear reactor, some extremely important restrictions follow from this condition.

Chapter 2

INTERACTION OF ELECTROMAGNETIC WAVES
WITH PLASMAS

2.1. BASIC CONCEPTS AND THE EQUATIONS OF
PLASMA ELECTRODYNAMICS

The excitation, propagation, and absorption of electromagnetic waves in plasmas, i.e., plasma electrodynamics, has developed extremely rapidly in recent years and occupies an important place in research on rf heating for thermonuclear experiments. The current state of this branch of physics has been fully reviewed and discussed in a series of detailed monographs [29–33]. There are also a number of books and reviews which discuss the latest advances in particular areas that are still under active investigation.

The problems of rf heating have to a substantial degree stimulated the development of plasma electrodynamics. It turns out that, in order to understand the processes taking place in a plasma during rf heating, it is often necessary to take extremely subtle effects into account. Meanwhile, not all branches of plasma electrodynamics are equally important for understanding the physics of rf heating. This circumstance makes it somewhat difficult to discuss the theory of rf plasma heating. On one hand, a detailed discussion of the appropriate part of plasma electrodynamics would greatly increase the size of this book. Any such discussion would also be fragmentary and would inevitably duplicate many of the standard monographs. On the other hand, some sort of introduction to plasma electrodynamics is nevertheless necessary. Without an introduction, it would be far more difficult to understand many topics in the physics of heating. Therefore, in the following we offer a sort of introduction to the electrodynamics of plasmas.

For coherence of exposition, we shall introduce an extended definition of the basic concepts and write down the basic equations (without any attempt at a detailed analysis). Subsequently, we shall draw a simplified, but physically as clear as possible, picture of the main wave processes in plasmas that are important for un-

27

derstanding the physics of heating and introduce (mostly without derivations) the principal formulas.

A macroscopic electromagnetic field in a plasma is described by the equations

$$\left.\begin{array}{l} \text{curl } \mathbf{E} = -\dfrac{1}{c}\dfrac{\partial \mathbf{B}}{\partial t}, \quad \text{div } \mathbf{E} = 4\pi(\rho + \rho_{ct}), \\[4mm] \text{curl } \mathbf{B} = \dfrac{1}{c}\dfrac{\partial \mathbf{E}}{\partial t} + \dfrac{4\pi}{c}(\mathbf{j} + \mathbf{j}_{ct}), \text{ div } \mathbf{B} = 0, \end{array}\right\} \tag{2.1}$$

where $\mathbf{E}$ is the electric field strength, $\mathbf{B}$ is the magnetic induction, $\mathbf{j}$ and ρ are the current and charge densities induced in the plasma, and $\mathbf{j}_{ct}$ and ρ_{ct} are the current and charge densities of the external sources.

This system of equations yields a relationship between the current density $\mathbf{j}$ and charge density ρ which expresses the conservation of charge:

$$\partial(\rho + \rho_{ct})/\partial t + \text{div}(\mathbf{j} + \mathbf{j}_{ct}) = 0.$$

In many cases it is convenient to use the electric displacement

$$\mathbf{D} = \mathbf{E} + 4\pi \int_{-\infty}^{t} \mathbf{j}(t')dt'. \tag{2.2}$$

instead of the current density $\mathbf{j}$.

In order to close the system of Eqs. (2.1), it is necessary to specify an equation of state, i.e., to establish the relationship between the current density $\mathbf{j}$ and the electric field strength of the wave $\mathbf{E}$. In the case of interest to us, a hot but not very dense plasma where the average potential energy of the particles is small compared to their average kinetic energy, this relationship is given in general by solving the kinetic equation

$$\frac{\partial f_j}{\partial t} + \mathbf{v}\frac{\partial f_j}{\partial \mathbf{r}} + \frac{e_j}{m_j}\left[\mathbf{E} + \frac{1}{c}(\mathbf{v}\times\mathbf{B})\right]\frac{\partial f_j}{\partial \mathbf{v}} = \frac{\delta f_j}{\delta t}, \tag{2.3}$$

where $f_j(\mathbf{v}, t)$ is the velocity distribution of particles of type j, e_j and m_j are the charge and mass of the particles, respectively, and $\delta f_j/\delta t$ is the collision integral which describes the changes in the distribution function owing to collisions. For a known distribution function the current density in the plasma is obviously defined as

$$\mathbf{j} = \sum_j e_j \int \mathbf{v} f_j(\mathbf{v})d\mathbf{v}. \tag{2.4}$$

In most plasma heating methods the wave frequency greatly exceeds the particle collision frequency so that the collisional term on the right-hand side of

Eq. (2.3) is usually a small correction. Even in the collisionless case, however, Eq. (2.3) is not solved in a general form, but an approximate solution, usually with several model assumptions, is employed to find the equation of state. Here the linear approximation for Eq. (2.3) is of fundamental importance, and only a correction (which is linear in the field) $f_1(v)$ to the equilibrium distribution function $f_0(v)$ is used to calculate the current. The kinetic equation linearized with respect to the wave field has the form

$$\frac{\partial f_{1j}}{\partial t} + v \frac{\partial f_{1j}}{\partial r} + \frac{e_j}{m_j c}(v \times B_0)\frac{\partial f_{1j}}{\partial v} = -\frac{e_j}{m_j}\left[E + \frac{1}{c}(v \times B) \right]\frac{\partial f_{0j}}{\partial v} + \left(\frac{\delta f_j}{\delta t}\right)_1, \qquad (2.5)$$

where B_0 is the unperturbed (by the wave) magnetic field in the device and E and B are the field strength and magnetic induction of the wave. (For brevity of notation here and in the following we shall omit the subscripts on the quantities E, B, j, and ρ referring to the wave.) The current density produced by the wave in the linear approximation is given by Eq. (2.4) with the exact distribution function $f(v)$ replaced by the linear approximation $f_1(v)$. It is clear from general considerations that in this case Eq. (2.4) has the form

$$j_\alpha = \Sigma_\beta \int_{-\infty}^{t} dt' \int \hat{\sigma}_{\alpha\beta}(t, t'; r, t')E_\beta(t', r')dr', \qquad (2.6)$$

so that the relationship between the current and field is nonlocal. The current at a given point in space at a given time is determined by the effect of the field over all space during all previous times. The physical reason for this nonlocalization, known as spatial and temporal dispersion, is fairly obvious. In fact, the current at a given point is proportional to the velocity and density of charged particles at that point. Since a plasma is a group of almost freely moving particles, however, their velocity at a given point at a given time depends on the action of the field on all previous parts of their trajectories; i.e., the current density has an integral dependence on the field.

Equation (2.6) is simpler for a spatially uniform, steady-state plasma. Then all points of space and all times are completely equivalent, so that the kernel of the integral in Eq. (2.6) depends on the differences of its arguments:

$$j_\alpha = \Sigma_\beta \int_{-\infty}^{z} dt' \int \hat{\sigma}_{\alpha\beta}(t-t'; r-r')E_\beta(t', r')dr'. \qquad (2.7)$$

Although real plasmas are always nonuniform, the model of a uniform plasma is extremely useful, since it allows us to classify the different branches of oscillations and waves which exist under the conditions and in the frequency ranges of interest to us. In the framework of this model it is natural to seek a plane wave solution of the system (2.1) and (2.7):

$$\mathbf{E}(\mathbf{r}, t) = \mathbf{E}(\omega, \mathbf{k}) \exp[i(\mathbf{k}\cdot\mathbf{r} - \omega t)], \quad \mathbf{B}(\mathbf{r}, t) = \mathbf{B}(\omega, \mathbf{k}) \exp[i(\mathbf{k}\cdot\mathbf{r} - \omega t)]. \quad (2.8)$$

Then, system (2.1) becomes a system of linear algebraic equations for the amplitudes $E_\alpha(\omega, \mathbf{k})$. Assuming that there are no external sources of the field and eliminating $\mathbf{B}(\omega, \mathbf{k})$, it is easy to obtain

$$k^2 \mathbf{E} - \mathbf{k}(\mathbf{k}\cdot\mathbf{E}) - (\omega^2/c^2)\mathbf{D} = 0, \quad (2.9)$$

where $D_\alpha = \sum_\beta \epsilon_{\alpha\beta} E_\beta$. The dielectric tensor $\epsilon_{\alpha\beta} \equiv \epsilon_{\alpha\beta}(\omega, \mathbf{k})$ is defined in accordance · with Eq. (2.2) by

$$\epsilon_{\alpha\beta} = \delta_{\alpha\beta} + (4\pi i/\omega)\sigma_{\alpha\beta}, \quad \sigma_{\alpha\beta} = \int_{-\infty}^{t} d\tau \int \hat{\sigma}_{\alpha\beta}(\tau; \mathbf{R}) \exp[-i(\mathbf{k}\cdot\mathbf{R} - \omega\tau)] d\mathbf{R}. \quad (2.10)$$

The condition that the system of uniform equations (2.9) for E_α have a nontrivial solution leads to the so-called dispersion relation

$$\det|k^2 \delta_{\alpha\beta} - k_\alpha k_\beta - (\omega^2/c^2)\epsilon_{\alpha\beta}| = 0, \quad (2.11)$$

which relates the frequency ω of the wave to its wave vector $\mathbf{k}$. It is possible to find the polarization of the wave (i.e., the relationship between the components E_α) from Eq. (2.9) by substituting the solution of the dispersion equation (2.11) into it. From the standpoint of the physics of rf plasma heating, the problems of greatest interest involve the propagation of waves excited by an antenna located outside the plasma. These problems correspond to solutions of the dispersion equation in the form $\mathbf{k}(\omega)$ with $\mathbf{k}$ determined for the frequency ω produced by the power source. Then the imaginary part of $\mathbf{k}(\omega)$ describes the spatial damping of the wave.

In its general form Eq. (2.11) is extremely complicated, and in order to classify the principal wave modes that are suitable for rf heating of plasmas in particular magnetic confinement systems, it is often convenient to consider the approximation of a cold, collisionless plasma. In this approximation, the plasma particles are treated as noninteracting and motionless in the absence of a wave. Here the current at a given point is clearly determined by the wave field at that point, so that the cou-

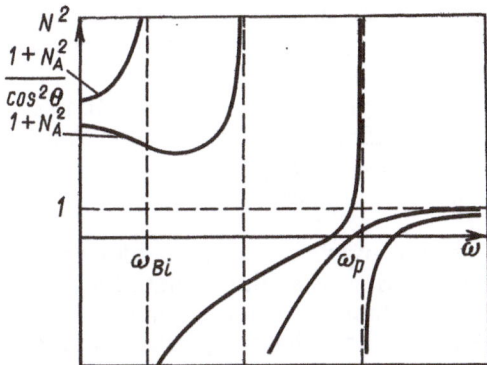

Fig. 2.1. The frequency dependence of the square of the refractive index in a cold plasma.

pling between the field and current becomes local (i.e., there is no spatial dispersion). The basic features of wave propagation and absorption in cold uniform plasmas which are important for rf heating are discussed below. Here, again, we shall only give a brief classification of the principal types of waves. In the cold plasma approximation, Eq. (2.11) takes the form

$$\tilde{A}_4 N^4 - \tilde{A}_2 N^2 + A_0 = 0, \tag{2.11a}$$

where $N = kc/\omega$ is the refractive index of the wave and the coefficients $\tilde{A}_4$, $\tilde{A}_2$, and $\tilde{A}_0$ are expressed fairly simply (Section 2.3) in terms of the components of the dielectric tensor and the angle θ between the vector k and the external magnetic field. It is clear from this equation that, in general, two types of waves, with different refractive indices, can exist in a cold plasma. These waves, however, propagate only for frequencies and plasma parameters such that $N^2 > 0$. The boundaries of the propagation region for the waves are determined by the conditions $\tilde{A}_0 = 0$ (the cutoff condition for one of the waves, corresponding to $N = 0$) and $\tilde{A}_4 = 0$ (the resonance condition, at which the refractive index of the wave becomes infinite).

At rf frequencies considerably higher than the characteristic frequencies of the plasma, the refractive indices of both waves are close to unity. At sufficiently low frequencies Eq. (2.11a) has the approximate solutions

$$\omega^2 = k^2 v_A^2, \qquad \omega^2 = k^2 v_A^2 \cos^2 \theta,$$

where $v_A \simeq \sqrt{B_0^2/(4\pi n_i m_i)}$ is the Alfvén speed, n_i is the plasma ion density, and m_i is the ion mass. The first of these solutions corresponds to fast magnetosonic waves and the second, to Alfvén waves. When the frequency ω is varied from zero to infinity, the coefficients $\tilde{A}_0$ and $\tilde{A}_4$ (for $\theta \neq 0$, $\pi/2$) both pass through zero three times. Thus, there are three cutoff frequencies and three resonance frequencies. The cutoff frequencies are independent of the angle of propagation of the wave. One cutoff occurs at the plasma frequency $\omega_{pe} = \sqrt{4\pi n_e e^2/m_e}$, where n_e, e, and m_e are the density, charge, and mass of the electron, respectively. The wave with a cutoff at the plasma frequency is referred to as the high-frequency branch of the ordinary wave. The second wave, with a resonance and two cutoffs at high frequencies, is called the extraordinary wave. As an illustration, Fig. 2.1 shows the function $N^2(\omega)$ for these waves at a propagation angle $\theta \neq 0, \pi/2$.

In general, waves are polarized elliptically, while the projection of the electric field of the wave in the direction of propagation is usually nonzero. Furthermore, in the neighborhood of a resonance the wave field is oriented predominantly along **k**.

The cold uniform plasma approximation provides a far from adequate description of the wave processes during rf heating. First of all, the spatial inhomogeneity of the plasma plays an important role under actual conditions. In most cases this inhomogeneity is weak in the sense that the plasma parameters vary little over distances on the order of the wavelength, so that the approximation of geometric optics is extremely productive as a natural generalization of the theory for a uniform plasma.

The thermal motion of the particles is also important, since in most cases it determines the way the electromagnetic energy is dissipated. Finally, nonlinear processes play a significant role in rf heating. The heating itself is the result of the nonlinear interaction of the waves with the plasma. Often in the first approximation it is possible to assume that this nonlinearity reduces merely to a smooth variation in the temperature of the particles and a perturbation in their distribution function, while both propagation and absorption of waves in a plasma with this particle distribution function are essentially linear. In many cases, however, this simple (so-called quasilinear) approach is inadequate. At high rf powers parametric instabilities develop in the plasma, generating waves whose frequencies differ considerably from that of the power source and creating turbulence.

The questions touched upon here have been analyzed in rigorous detail in an extensive literature [7, 29–35]. We shall not reproduce this analysis here. Instead, in the following sections we shall discuss the qualitative physics of the most important phenomena involved in plasma heating and introduce some important computational formulas.

2.2. THE DIELECTRIC TENSOR OF COLD PLASMAS

As noted in the preceding section, the dielectric tensor is generally obtained by solving the kinetic equation (2.5), calculating the current density according to Eq. (2.4), and using Eqs. (2.10). For a cold collisionless plasma, where the charged particles are at rest in the absence of a wave, this calculation can be done on the basis of a model with independent particles. Let the plasma lie in an external magnetic field B_0 directed along the Z axis. We shall consider the motion along $\mathbf{B}_0$ of a charged particle acted on by a plane electromagnetic wave:

$$m_j \frac{du_{zj}}{dt} = e_j \left\{ E_z + \frac{1}{c} [\mathbf{u}(\mathbf{B}_0 + \mathbf{B})]_z \right\} \exp\left[i(\mathbf{k} \cdot \mathbf{r} - \omega t) \right]. \qquad (2.12)$$

Here u_j is the directed velocity of a particle acted on by the wave. In the linear approximation the second term in the curly brackets should be dropped and the variation of r in the exponent neglected. Then, assuming that the wave field is turned on adiabatically slowly at an infinitely removed time (i.e., assuming that ω has a small positive imaginary part), we find the velocity of a particle of a given type j to be

$$u_{zj} = [ie_j/(m_j\omega)]\, E_z \exp[i(\mathbf{k\cdot r} - \omega t)]. \tag{2.13}$$

The current density produced by these particles is obtained by multiplying u_{zj} by the charge e_j and particle density n_{0j}. The total current density in the plasma is found by summing the contributions from particles of different types:

$$j_z = \sum_j e_j n_{0j} u_{zj} = i \sum_j \frac{e_j^2 n_{0j}}{m_j\,\omega}\, E_z \exp[i(\mathbf{k\cdot r} - \omega t)]. \tag{2.14}$$

From this it is clear that the corresponding components of the conductivity tensor σ_{zz}, σ_{xz}, and σ_{yz} have the form

$$\sigma_{zz} = \frac{i}{4\pi} \sum_j \frac{\omega_{pj}^2}{\omega}, \quad \sigma_{xz} = \sigma_{yz} = 0, \tag{2.15}$$

where $\omega_{pj} = \sqrt{4\pi e_j^2 n_{0j}/m_j}$ is the plasma frequency of j-type particles.

Using Eq. (2.10), we find the longitudinal component of the dielectric tensor to be

$$\epsilon_{zz} \equiv \eta = 1 - \sum_j \omega_{pj}^2/\omega^2; \quad \epsilon_{zx} = \epsilon_{zy} = 0. \tag{2.16}$$

In order to calculate the transverse components of the conductivity tensor we examine the currents produced by circularly polarized fields $E^\pm = (\mathbf{e}_x \pm i\mathbf{e}_y)E \times \exp[i(\mathbf{k\cdot r} - \omega t)]$ with frequency ω, where $\mathbf{e}_{x,y}$ are the unit vectors along the X and Y axes, respectively. It is easy to see that under the influence of such a field (with adiabatic turn on) a charged particle also will rotate in a circle in the same direction as the electric field vector. The velocity of the particle, however, will be different in different rotation directions. When E rotates counter to the cyclotron rotation of the particles, the centripetal force $F_c = m\omega u$ (here $u = \sqrt{u_x^2 + u_y^2}$) on a particle will be equal to the difference between the "electric" force $F_E = |e|E$ and the Lorentz force $F_L = (|e|/c)uB_0 = m\omega_B|u$ (Fig. 2.2a), so that $u = (e/m)E/(\omega + \omega_B|)$. When E rotates in the opposite direction, the centripetal force $m\omega u$ and the Lorentz force $m\omega_B|u$ are in the same direction (Fig. 2.2, b) and the electric force $|e|E$ must either augment the effect of the Lorentz force (for $\omega > \omega_B|$) or reduce it (for $\omega < \omega_B|$). As a result, the velocity u of rotation of the particle will be given by the ratio $|e|\,|E|/[m(\omega - \omega_B|)]$. In vector form with complex notation this ratio can be written in the form

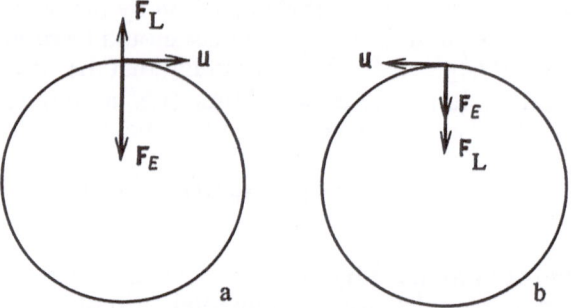

Fig. 2.2. The motion of a particle in a circularly polarized wave field with the vector **E** rotating opposite to (a) and in the same sense (b) as the cyclotron rotation of the particle ($\omega > | \omega B |$).

$$\mathbf{u}^\pm = i e\, \mathbf{E}^\pm / [m (\omega \pm \omega_B)], \tag{2.17}$$

where $\omega_B = e B_0 / (mc)$ is the cyclotron frequency. This formula can easily be derived directly from the equations of motion.

Noting that the combination $\mathbf{E}^+ + \mathbf{E}^-$ corresponds to a linearly polarized field $E_x = 2E$, $E_y = 0$, we find the current density for a wave polarized linearly along the X axis to be

$$j_x = \sigma_{xx} E_x, \quad j_y = \sigma_{yx} E_x,$$

where

$$\sigma_{xx} = \frac{i}{4\pi} \sum_j \frac{\omega_{pj}^2 \omega}{\omega^2 - \omega_{Bj}^2}, \quad \sigma_{yx} = \frac{1}{4\pi} \sum_j \frac{\omega_{pj}^2 \omega_{Bj}}{\omega^2 - \omega_{Bj}^2}. \tag{2.18}$$

The corresponding components of the dielectric tensor are

$$\epsilon_{xx} = \epsilon_{yy} \equiv \epsilon = 1 - \sum_j \omega_{pj}^2 / (\omega^2 - \omega_{Bj}^2), \tag{2.19}$$

$$\epsilon_{xy} = -\epsilon_{yx} \equiv ig = -i \sum_j \omega_{pj}^2 \omega_{Bj} / \omega (\omega^2 - \omega_{Bj}^2). \tag{2.20}$$

In a cold plasma, therefore, the dielectric tensor has the form

$$\overset{\smile}{e} = \begin{pmatrix} \epsilon & ig & 0 \\ -ig & \epsilon & 0 \\ 0 & 0 & \eta \end{pmatrix}. \tag{2.21}$$

We note that in this approximation the current produced by the wave is purely reactive and there is no energy exchange between the wave field and the plasma on the average over the period of the oscillations. In fact, the electromagnetic energy absorbed in the medium during a period of the oscillations is given by

$$p_E = \overline{\text{Re } j \cdot \text{Re } E} = \frac{1}{4} \sum_\alpha (j_\alpha E_\alpha^* + j_\alpha^* E_\alpha).$$

Using the equality $j_\alpha = \sum_\beta \sigma_{\alpha\beta} E_\beta$ and changing the indices in the second term, we find

$$p_E = \frac{1}{2} \sum_{\alpha,\beta} \sigma'_{\alpha\beta} E_\alpha^* E_\beta = \frac{\omega}{8\pi} \sum_{\alpha,\beta} \epsilon''_{\alpha\beta} E_\alpha^* E_\beta, \qquad (2.22)$$

where $\sigma_{\alpha\beta}' = (\sigma_{\alpha\beta} + \sigma_{\beta\alpha}^*)/2$ is the Hermitian and $i\epsilon_{\alpha\beta}'' = (\epsilon_{\alpha\beta} - \epsilon_{\beta\alpha}^*)/2$ is the anti-Hermitian part of the tensors $\sigma_{\alpha\beta}$ and $\epsilon_{\alpha\beta}$, respectively. The tensor (2.21) is Hermitian, so that there is no absorption of wave energy in a cold collisionless plasma. Naturally, this is a consequence of the crudeness of the model. Thus, if we introduce a phenomenological collision term in Eq. (2.12) in the form of an effective frictional force $m_j u_{zj} \nu_j$ (where ν_j is the collision frequency), then

$$\eta = 1 - \sum_j \frac{\omega_{pj}^2}{\omega(\omega + i\nu_j)} \simeq 1 - \sum_j \left(\frac{\omega_{pj}^2}{\omega^2} - i \frac{\omega_{pj}^2 \nu_j}{\omega^3} \right). \qquad (2.23)$$

It is clear that in this case the current has an "active" component and the anti-Hermitian part of $\epsilon_{\alpha\beta}$ is nonzero.

2.3. OSCILLATIONS AND WAVES IN COLD PLASMAS

Using the dielectric tensor of the form (2.21) in Eq. (2.9) and introducing the refractive index vector $N = kc/\omega$, we obtain

$$
\left.
\begin{aligned}
(\epsilon - N_z^2) E_x + ig E_y + N_x N_z E_z &= 0; \\[2mm]
(\epsilon - N_x^2 - N_z^2) E_y - ig E_x &= 0, \\[2mm]
(\eta - N_x^2) E_z + N_z N_x E_x &= 0,
\end{aligned}
\right\} \qquad (2.24)
$$

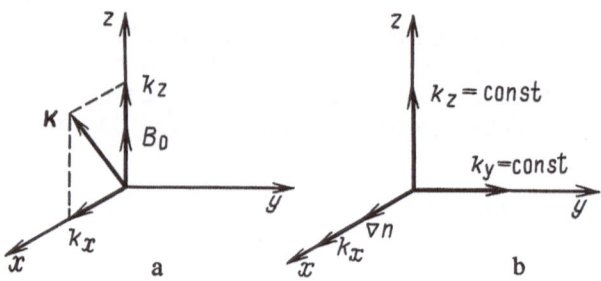

Fig. 2.3. The coordinate systems used to study the propagation of waves in uniform plasmas (a) and in plasmas with a density inhomogeneity (b).

where it is assumed that the X axis lies in the plane determined by the vectors $\mathbf{B_0}$ and $\mathbf{N}$ (Fig. 2.3a). The dispersion relation (2.11) for this case can be written in the form

$$\epsilon N_x^4 - A_2 N_x^2 + A_0 = 0, \qquad (2.25)$$

where $A_2 = (\epsilon + \eta)(\epsilon - N_z^2) - g^2$; and $A_0 = \eta[(\epsilon - N_z^2)^2 - g^2]$.

Waves in a cold plasma can be classified in different ways. It is often convenient to consider the dependence of the refractive index $N = \sqrt{N_x^2 + N_z^2}$ on the angle θ between the vector $\mathbf{N}$ and the external magnetic field $\mathbf{B_0}$. Setting $N_z^2 = N^2 \times \cos^2\theta$ and $N_x^2 = N^2 \sin^2\theta$ in Eq. (2.25) we obtain a biquadratic equation for N of the form (2.11a), so that in general there are two independent solutions of Eq. (2.25) for $N^2(\theta)$. For a fixed ion composition in the plasma, the behavior of the functions $N(\theta)$ is determined by the dimensionless parameters ω_{pe}^2/ω^2 and ω_{Be}^2/ω^2. It turns out [32] that in the parameter space (ω_{pe}^2/ω^2, ω_{Be}^2/ω^2) there are a number of regions (13 when there is only one type of ion in the plasma) within which the topology of the curves $N(\theta)$ plotted in polar coordinates (N, θ) does not change. By displaying curves of this type for each of these regions, it is possible to visualize the behavior of the refractive index over the entire range of variation of the plasma parameters.

It will be apparent later, however, that this approach is not very convenient for examining boundary problems in nonuniform plasmas. The two solutions of Eq. (2.25) of the form $N_x^2(N_z^2)$ generalize more simply to the case of a nonuniform plasma when the variable parameter is taken to be a component of the refractive index along the magnetic field, rather than the angle θ of propagation of the wave. Both approaches are equivalent, and the choice between the two is determined by considerations of convenience.

A detailed analysis of the solutions of Eq. (2.25) can be found elsewhere [29, 30, 32, 36]. Here we shall consider a number of simple, but important examples.

We first consider the polarization of waves in plasmas. Electromagnetic waves in dielectrics and in a vacuum are purely transverse ($\mathbf{k \cdot E} = 0$, $\mathbf{k \cdot B} = 0$) because of the absence of space charge. However, in a plasma, which is made up of essentially freely moving charged particles, local fluctuations in the electron and ion densities relative to one another are entirely possible, so that space charge can also develop. On the average over long periods of time, of course, the plasma remains spatially quasineutral.

Space charge causes the electromagnetic field of the wave to develop a component parallel to the wave vector. As an example we consider the case $\mathbf{k} \parallel \mathbf{B}_0$ (i.e., $N_x = 0$). Then it is easy to show that the system of Eqs. (2.24) allows the solution

$$\left. \begin{array}{l} E_x = 0, \quad E_y = 0, \quad \eta E_z = 0, \\ \eta = 1 - (\omega_{pe}^2 + \omega_{pi}^2)/\omega^2 = 0, \end{array} \right\} \qquad (2.26)$$

which describes the field of a space charge wave propagating along an external magnetic field. In the cold plasma approximation the frequency of this wave is identical to the plasma frequency, regardless of the refractive index N_z. This result is also easily obtained from a simple physical picture. Let a plane layer of plasma with zero net charge lie perpendicular to an external magnetic field. If the electrons in the layer are displaced a distance ξ relative to the ions (along the field), then a surface charge density $\rho_s = en_e\xi$ develops and produces an electric field $E = -4\pi\rho_s$ which will tend to return the electrons to their original position. Neglecting the ion motion, we can write

$$d^2 \xi / dt^2 = (e/m_e)E = -\omega_p^2 \xi.$$

This shows that free oscillations of the electrons in such a layer also take place at the plasma frequency. Taking the ions' motion into account yields only a minor change in the frequency of the oscillations, i.e., $\omega = \sqrt{\omega_{pe}^2 + \omega_{pi}^2} \approx \omega_{pe}$.

We shall now try to examine the space charge oscillations in a plane perpendicular to the magnetic field (with the plasma layer parallel to the YOZ plane) in the same way. Here a restorative force will arise because of the action of the magnetic field as well as of the space charge electric field:

$$\frac{d^2 \xi}{dt^2} = -\omega_{pe}^2 \xi + \omega_{Be} v_y, \qquad \frac{dv_y}{dt} = -\omega_{Be}\frac{d\xi}{dt}, \qquad (2.27)$$

so that the motion of the electrons along the X axis (with motionless ions) will obey the equation

$$d^2 \xi / dt^2 = -(\omega_{pe}^2 + \omega_{Be}^2) \xi.$$

The characteristic oscillations of the electron space charge transverse to the magnetic field will, therefore, have a frequency $\omega = \omega_{uh} \equiv \sqrt{\omega_{pe}^2 + \omega_{Be}^2}$, which is known as the upper hybrid resonance frequency. We note an important difference between the oscillations parallel and perpendicular to the magnetic field: an inhomogeneity in the space charge along the magnetic field (along the Z axis) leads to motion of the charges (and, therefore, to an electric field) only along the Z axis, while according to Eq. (2.27), an inhomogeneity in the space charge along the X axis leads to motion of charges (and, therefore, to a current) along the Y axis. A current $j_y \neq 0$ usually should generate a rotational (transverse) field E_y so that this picture is not entirely correct. In order to settle this question, it is convenient to proceed directly from Eqs. (2.24), which for $N_z = 0$ and $E_z = 0$ take the form

$$\left. \begin{array}{l} \epsilon E_x + ig\, E_y = 0, \\[2mm] (\epsilon - N_x^2) E_y - ig\, E_x = 0. \end{array} \right\} \tag{2.28}$$

This yields the dispersion equation

$$\epsilon N_x^2 = \epsilon^2 - g^2 \tag{2.29}$$

and, for the ratio of the components E_x and E_y,

$$E_x / E_y = -ig / \epsilon. \tag{2.30}$$

In this case, therefore, a wave with "mixed" polarization can propagate in the plasma at all frequencies satisfying $N_x^2 > 0$. It is called the extraordinary wave. Since $\epsilon \to 0$ when $\omega \to \omega_{uh}$ in the high-frequency limit when the ion motion can be neglected, we conclude that for $\omega = \omega_{uh}$ the polarization of the extraordinary wave, as assumed before, is purely longitudinal ($E_y/E_x = 0$). It follows from Eq. (2.29), however, that the refractive index of the extraordinary wave goes to infinity as $\omega \to \omega_{uh}$, so that this frequency is the frequency of a unique resonance of the wave (known as the upper hybrid resonance).

In order to illustrate more clearly the physical reason for shortening of the wavelength at a resonance, we return to the wave equations (2.1), rewritten in the form

$$\nabla \times \nabla \times \mathbf{E} = -\frac{1}{c^2} \frac{\partial^2 \mathbf{D}}{\partial t^2} = -\frac{1}{c^2} \frac{\partial^2 \mathbf{E}}{\partial t^2} - \frac{4\pi}{c^2} \frac{\partial \mathbf{j}}{\partial t}. \tag{2.31}$$

The projection of this equation on the Y axis is of interest to us and in this case has the form

$$\nabla^2 E_y = -k_x^2 E_y = \frac{1}{c^2} \frac{\partial^2 E_y}{\partial t^2} + \frac{4\pi}{c} \frac{\partial j_y}{\partial t};$$

i.e., it is clear that the difference of the refractive index from unity (the difference between k_x and ω/c) is caused by a current induced in the plasma by the wave. The wave is slowed down strongly when the current in the plasma is much greater than the displacement current. Clearly, as the frequency of the eigenmodes is approached, j_y increases resonantly because of the above-noted coupling between the motion of particles in the x and y directions [see Eq. (2.27)]. This is also the reason for the resonant increase in the wave vector of these oscillations.

It is evident from Eq. (2.31) that for finite $\sigma_{\alpha\beta}$ in the case of a resonance $k \to \infty$, a wave must be approximately longitudinally polarized ($\mathbf{E} = -\nabla\varphi$) for arbitrary propagation angles. In fact, only then does the left-hand side of Eq. (2.31) (with the spatial derivatives) not go to infinity as $k \to \infty$. This consideration makes it possible to obtain the resonance conditions in the general case. Substituting $\mathbf{E} = -\nabla\varphi$ in the equation

$$\text{div } \mathbf{D} = \text{div } \check{\epsilon}\mathbf{E} = 0, \tag{2.32}$$

which is a consequence of Eq. (2.31), we obtain

$$\sin\theta\cos\theta\,(\epsilon_{xz} + \epsilon_{zx}) + \sin^2\theta\,\epsilon_{xx} + \cos^2\theta\,\epsilon_{zz} = 0, \tag{2.33}$$

where $\sin^2\theta = k_x^2/k^2$, $\cos^2\theta = k_z^2/k^2$. In the cold plasma approximation, where $\epsilon_{\alpha\beta}$ is given by Eq. (2.21), Eq. (2.33) takes the form

$$\epsilon\sin^2\theta + \eta\cos^2\theta = 0. \tag{2.33a}$$

We shall discuss the effects associated with the hybrid resonances in more detail later. But for now, we return to the case of wave propagation perpendicular to the external magnetic field. Equations (2.28) and (2.29) describe a wave whose electric field lies in a plane perpendicular to $\mathbf{B}_0$. Setting $E_x = E_y = 0$ and $N_z = 0$ in Eqs. (2.24), we obtain a dispersion relation for the ordinary wave whose electric field is polarized along $\mathbf{B}_0$:

$$N_x^2 = \eta. \tag{2.34}$$

The ordinary wave causes motion of charges only along the vector $\mathbf{B}_0$, so that the magnetic field has no effect on its propagation when $\theta = \pi/2$. This wave has a purely transverse polarization ($\mathbf{k}\cdot\mathbf{E} = 0$) and, therefore, experiences no resonances.

Purely transverse waves also exist for longitudinal propagation $\theta = 0$. For $E_z = 0$ and $N_x = 0$ the system of equations (2.24) reduces to the form

$$(\epsilon - N_z^2)\,E_x + ig\,E_y = 0,$$
$$(\epsilon - N_z^2)\,E_y - ig\,E_x = 0.$$

Adding the first equation to the products of the second by $+i$ and $-i$, we obtain

$$\left.\begin{array}{l}(\epsilon - N_z^2 + g)\,(E_x + iE_y) = 0, \\[2mm] (\epsilon - N_z^2 - g)\,(E_x - iE_y) = 0.\end{array}\right\} \tag{2.35}$$

This implies that when $\theta = 0$, two other circularly polarized waves can propagate in the plasma ($E_x \pm iE_y = 0$) besides the longitudinal wave described by Eq. (2.26). Their refractive indices are given by

$$N_z^2 = \epsilon \pm g = 1 - \frac{\omega_{pe}^2}{\omega(\omega \pm \omega_{Be})} - \frac{\omega_{pi}^2}{\omega(\omega \pm \omega_{Bi})}. \tag{2.36}$$

The wave whose electric field **E** rotates in the direction of the electron's cyclotron rotation experiences a resonance at the electron cyclotron frequency ($\omega = |\omega_{Be}|$) and the wave rotating in the opposite direction has a resonance at the ion cyclotron frequency ($\omega = \omega_{Bi}$). The physics of these resonances is quite obvious. They are associated with the cyclotron resonances of "isolated" particles and, in this sense, they are very different from the resonances described by Eq. (2.33) which are determined by "collective" properties of the charges (i.e., the field that develops because of charge separation). The field is not longitudinal at the cyclotron resonance because the components ϵ and g of the dielectric tensor go to infinity.

It should be noted that the cyclotron resonance with longitudinal propagation ($\theta = 0$) is distinct. When $\theta \neq 0$, the refractive index of the waves has no singularities at the electron or ion cyclotron frequencies, although the perpendicular components of the tensor $\check{\epsilon}$ go to infinity at that point. This is also evident, in particular, from Eq. (2.29) when $\theta = \pi/2$. The reason for the lack of a resonance during oblique propagation can be understood from the following simple considerations. Equation (2.9) shows that the component of the displacement vector $\mathbf{D} = \mathbf{E} + (4\pi i/\omega)\mathbf{j}$ along the wave vector **k** must be zero (i.e., $\mathbf{k} \cdot \mathbf{D} = 0$). However, when the electric field of the wave contains a resonant component rotating in the direction of the cyclotron rotation of the particles, because of the resonant particles a "rotating" current $\mathbf{j}_\perp$ develops in the plasma [Eq. (2.17)] with an amplitude which approaches infinity at the resonance, so that the component of **D** perpendicular to $\mathbf{B}_0$ also goes to infinity. Then the requirement that $\mathbf{k} \cdot \mathbf{D} = 0$ can be satisfied in the presence of a resonant component in the electric field only for purely longitudinal propagation of the wave ($\theta = 0$). For oblique propagation, the resonant component of the electric field goes to zero (i.e., at the electron cyclotron resonance the vector $\mathbf{E}_\perp$ rotates in the ion direction, and vice versa). In the special case of perpendicular propagation, Eq. (2.30) gives exactly the same picture for the wave polarization.

2.4. WAVE PROPAGATION IN TOKAMAKS IN THE APPROXIMATION OF GEOMETRIC OPTICS

The magnetic field geometry and plasma nonuniformity in tokamaks are extremely complex, so that it is rather difficult to obtain a rigorous solution of the wave propagation problem. The problem can be greatly simplified if the relatively small poloidal magnetic field is neglected and if wave propagation near the equatorial plane of the torus is considered. Then it is possible to restrict oneself to the model of a plane plasma layer $-a < x < a$ located in a "toroidal" magnetic field $B_0 = B(0)R_0/(R_0 + x)$ directed along the Z axis. In this model the point $x = a$ corresponds to the outer, and $x = -a$ to the inner, plasma boundary in the equatorial plane of the torus. The plasma density in this layer is assumed to be uniform in the Y and Z directions, with a maximum at $x = 0$ corresponding to its value on the magnetic axis of the tokamak and falling toward the boundaries at $x = \pm a$.

In most cases of practical interest, the plasma minor radius a, which determines the scale of the plasma nonuniformity, is considerably greater than the wavelength of the electromagnetic waves used for heating. Under these conditions it is natural to use the well-known approximation of geometric optics for describing the propagation of the waves. In this approximation the wave field is represented in the form

$$\mathbf{E}(\mathbf{r}, t) = \mathbf{E}(x)\exp\left[ik_y y + ik_z z + i\int^x k_x(x')dx' - i\omega t\right], \tag{2.37}$$

where the wave amplitude $\mathbf{E}(x)$ and the projection x of the geometric-optical wave vector $k_x(x)$ are regarded as smooth functions which vary little over distances on the order of the wavelength, so that the derivatives of $\mathbf{E}(\mathbf{r}, t)$ with respect to x can be neglected in the zeroth approximation. In this approximation all of the equations in Sections 2.2 and 2.3 retain their form when only the exponent is differentiated if we take ω_p and ω_B to denote their local values. One insignificant difference arises solely because of a different choice of the coordinate axes (in Section 2.3 the Y axis was directed perpendicular to the vectors $\mathbf{k}$ and $\mathbf{B}_0$; see Fig. 2.3b). Therefore, in Eqs. (2.25), (2.29), and (2.34) of Section 2.3 N_x^2 should be replaced by $N_\perp^2 = N_x^2 + N_y^2$. Then the dispersion relation (2.25) determines the local value of $N_x(x)$ in terms of specified values of N_y and N_z on the plasma boundary. After $N_x(x)$ is substituted into Eqs. (2.24) and the coordinate axes are rotated appropriately, these equations will determine the polarization of the wave [i.e., the spatial orientation of $\mathbf{E}(x)$] at any point, just as in a uniform plasma. The geometric-optical wave amplitude $|\mathbf{E}(x)|$ is not given by the zeroth approximation. In order to find it, one must examine the first approximation of geometric optics, i.e., take the first derivatives $d\mathbf{E}(x)/dx$ and $dk_x(x)/dx$ into account, treating them as small corrections. A detailed procedure for finding the first and higher approximations of geometric optics has been given by Ginzburg [30]. We shall not discuss it here, since the wave propagation behavior of interest to us can be established using the zeroth approximation alone.

    We consider only those waves with frequencies high enough that the terms in $\epsilon_{\alpha\beta}$ owing to the ion motion can be neglected. We begin with the case $N_z = 0$. For the ordinary wave, according to Eq. (2.34),

$$N_x^2 = 1 - \omega_{pe}^2(x)/\omega^2 - N_y^2. \tag{2.38}$$

Evidently, $N_x^2 > 0$ only in the plasma region where $\omega_{pe}^2(x)/\omega^2 < 1 - N_y^2$. Physically, it is clear that at the point x_0 such that $N_x^2 = 0$ the wave is reflected ("cut off") and its field is attenuated exponentially with depth in the opacity region $\omega_{pe}^2(x)/\omega^2 > 1 - N_y^2$ as $\mathbf{E}(\mathbf{r}) \propto \exp[-(\omega/c)\int_{x_0}^x N_x(x')dx']$. From a formal standpoint, the approximation of geometric optics breaks down near x_0, and the wave equation must be examined more carefully. Here we limit ourselves to physical arguments.

    One must not confuse a reflection point in a real nonuniform plasma ($N_x = 0$) with the cutoff point ($N^2 = N_x^2 + N_y^2 + N_z^2 = 0$) used in classifying the waves in the theory of uniform plasmas [1, 5, 10]. In general, they coincide only when the wave propagates along the density gradient of the plasma. Then the ordinary wave may reach a layer with a maximum plasma density corresponding to the condition $\omega_{pe}^2/\omega^2 = 1$.

For the extraordinary wave at $N_z^2 = 0$, it is possible to write

$$N_x^2(x) = \frac{[\omega(\omega - \omega_{Be}) - \omega_{pe}^2][\omega(\omega + \omega_{Be}) - \omega_{pe}^2]}{\omega^2(\omega^2 - \omega_{uh}^2)} - N_y^2 \qquad (2.39)$$

in accordance with Eq. (2.29). This shows that the location of the reflection points depends strongly on the magnetic field. For $N_y = 0$ they are given by

$$\omega_{pe}^2/\omega^2 = 1 - \omega_{Be}(x)/\omega, \qquad (2.40)$$

$$\omega_{pe}^2/\omega^2 = 1 + \omega_{Be}(x)/\omega. \qquad (2.41)$$

At the hybrid resonance point ($\omega^2 = \omega_{uh}^2 = \omega_{pe}^2(x) + \omega_{Be}^2(x)$), N_x^2 goes to infinity. For concreteness, let the cyclotron resonance condition $\omega = |\omega_{Be}(x)|$ be satisfied at the center $x = 0$ of the plasma layer. Then in a dense plasma with $\omega_{pe}^2(0)/\omega^2 > 2$ Eq. (2.40) is satisfied to the left and right of $x = 0$ and the cyclotron resonance is located in the opacity region (see Fig. 2.4.) The condition (2.41) is satisfied at some point $x_0 > 0$ (for $\omega > |\omega_{Be}|$) and the entire plasma region between x_0 and the hybrid resonance point is also opaque. Thus, in a dense plasma the hybrid and electron cyclotron resonances are inaccessible for waves excited at the plasma edge with $N_y = 0$. (Strictly speaking, this conclusion refers only to waves propagating near the equatorial plane of a torus.)

As $\omega_{pe}^2(0)/\omega^2$ is reduced, the two cutoff points (2.40) come closer together and when

$$\frac{\omega_{pe}^2(0)}{\omega^2} < 1 + \sqrt{1 - \frac{a_0^2}{(4R^2)}} \left(a_0^{-2} = -\frac{1}{2} \frac{\partial^2}{\partial x^2} \frac{\omega_{pe}^2}{\omega^2} \bigg|_{x=0} \right)$$

they disappear. Then a wave excited at $x = -a$ (the inner edge of the torus) can propagate all the way to the hybrid resonance layer, freely passing through the plane of the electron cyclotron resonance (Fig. 2.5).

It is now easy to see how the N_x^2 curve behaves when $N_y^2 \neq 0$. In accordance with Eq. (2.29), it is sufficient to imagine lowering the curves along the ordinate by N_y^2 in Figs. 2.4 and 2.5. The cutoff points are then shifted toward the plasma edge (toward lower density) while the location of the hybrid resonance is unchanged.

As in a uniform plasma, the refractive index of the wave does not have a singularity at the point $\omega - |\omega_{Be}(x)| = 0$. The physical reason that the refractive index goes to infinity at the hybrid resonance $\omega = \omega_{uh}$ is (Section 2.3) resonant pumping of space charge eigenmodes. Then, as follows from Eq. (2.30), the electric field component E_x also increases resonantly as the hybrid resonance point x_r is approached, i.e.,

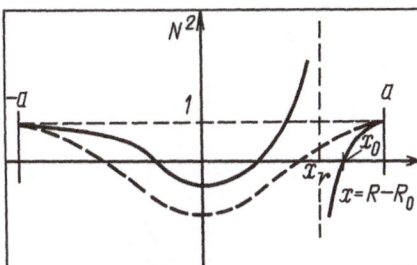

Fig. 2.4. The distributions of the square of the refractive index for ordinary (- - -) and extraordinary (——) waves over the cross section of a plasma column.

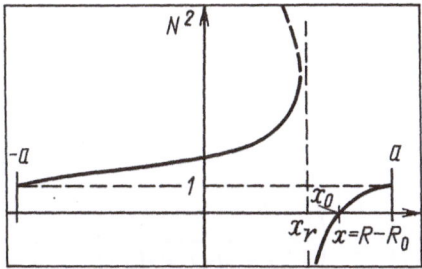

Fig. 2.5. The dependence of the refractive index of an extraordinary wave on the distance from the symmetry axis of a torus: —— cold plasma approximation; - - - plasma wave caused by thermal motion of the particles.

$$E_x \simeq -\mathrm{i}g\,(x_r)\,a_r E_y\,(x_r)/(x-x_r),$$

where it is assumed that $\epsilon \simeq (x-x_r)/a_r$ in the neighborhood of the resonance and for simplicity that $N_y = 0$. Thus, from this point of view it is natural to assume that the wave is longitudinally polarized at the hybrid resonance because the longitudinal component E_x goes to infinity, rather than because the transverse component E_y goes to zero. This assumption is confirmed by a rigorous analysis of the wave equations [30].

The conclusion that E_x goes to infinity at $x = x_r$ is obviously related to the complete neglect of the electron's thermal motion and of dissipation of the wave energy in the plasma. As noted above, dissipation leads to the appearance of an anti-Hermitian part in $\epsilon_{\alpha\beta}$. Thus, when collisions occur, we can write, by analogy with Eq. (2.23), $\epsilon \simeq (x-x_r)/a_r + \mathrm{i}\nu_{\mathrm{eff}}/\omega$ (where ν_{eff} is proportional to the collision frequency), so that when $x = x_r$ the wave field remains finite $[E_x(x_r) \propto \omega/\nu_{\mathrm{eff}}]$. The power Q_r absorbed per unit surface area near the resonance layer is roughly

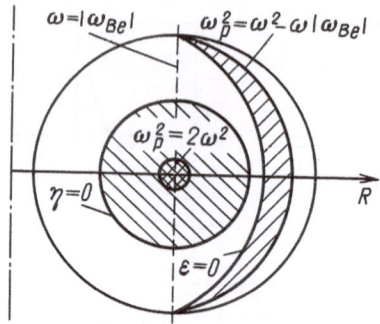

Fig. 2.6. The locations of the cutoffs and resonances in the minor cross section of a plasma column for $\omega \approx |\omega_{Be}|$. (The opacity regions for the waves are shaded.)

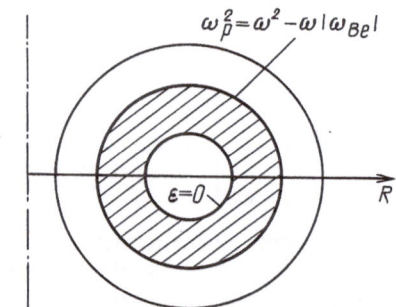

Fig. 2.7. The locations of the cutoffs and resonances in a plasma column for $\omega > |\omega_{Be}|$.

$$Q_r = \frac{1}{2}\int \sigma' E_x^* E_x dx = \frac{v_{\text{eff}}}{8\pi}\int_{-\infty}^{\infty} \frac{g^2(x_r)a_r^2|E_y(x_r)|^2}{(x-x_r)^2 + a_r^2 v_{\text{eff}}^2/\omega^2}dx.$$

Transforming to the new variable $\xi = [(x-x_r)/a_r]/(\omega/v_{\text{eff}})$ in this integral, we obtain

$$Q_r = Q_0 \int_{-\infty}^{\infty} \frac{d\xi}{1+\xi^2} = \pi Q_0, \tag{2.42}$$

where $Q_0 = (\omega a g^2(x_r)|E_y(x_r)|^2)/8\pi$.

This quantity is evidently independent of v_{eff}/ω, so that absorption of the wave remains finite even in the limit of vanishingly small dissipation. We can obtain the same result by assuming that $v_{\text{eff}} = 0$ while including a finite imaginary part of the

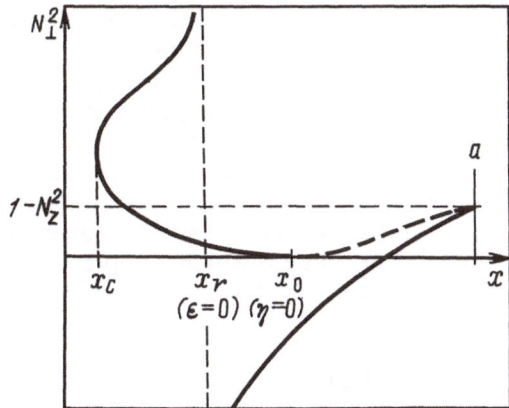

Fig. 2.8. The distribution of the square of the perpendicular refractive index $N_\perp^2$ of waves in the equatorial plane of a tokamak for $N_z^2 = |\omega_{Be}|/\omega + |\omega_{Be}|$.

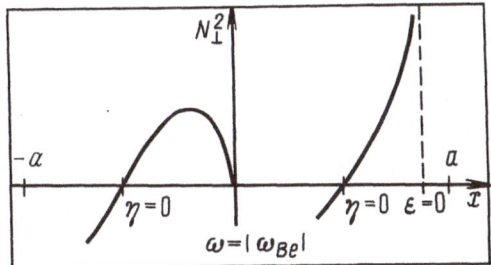

Fig 2.9. The distribution of the square of the refractive index $N_\perp^2$ for $N_z^2 \gg 1$.

frequency ω corresponding to adiabatic turning on of the field as $t \to -\infty$ and then letting it go to zero. Here the finite absorption can be interpreted as a loss of energy by the "external" electric field in pumping longitudinal space charge oscillations.

This type of resonant absorption of the wave energy has been studied in detail by many authors [30, 37, 38]. Its efficiency depends strongly on the scale length of the plasma inhomogeneities. In the case of interest to us – a weakly nonuniform plasma with $\omega a_r/c \gg 1$ where the opacity barrier between the points x_0 and x_r (Figs. 2.4 and 2.5) is usually very deep – a wave incident on the hybrid layer from $x < x_r$ (i.e., from the "accessible" side) is almost completely absorbed. A wave incident on the layer from the vacuum on the weak magnetic field side is reflected from the opacity region and absorption is exponentially small.

These results provide an idea of wave propagation behavior when $N_z^2 = 0$ in the equatorial plane. In general, these results cannot be extended to regions away from the equatorial plane because of the essentially two-dimensional character of the plasma inhomogeneity in these regions. An intuitive idea of wave propagation can be obtained, however, if we map the characteristics corresponding to the hybrid resonance $\omega = \omega_{uh}(\mathbf{r})$, the cutoffs for the extraordinary wave (2.40) and (2.41), the cutoff for the ordinary wave $\omega_{pe}^2(r) = \omega^2$, and the cyclotron resonance $|\omega_{Be}(r)| = \omega$ in the minor cross section of the torus and shade the region $N_\perp^2 < 0$ (Figs. 2.6 and 2.7). Figure 2.6 corresponds to the case in which the condition $\omega = |\omega_{Be}(x)|$ can occur within the plasma. Then the propagation region for an extraordinary wave launched from the weak magnetic field side is separated from the hybrid resonance curve by an opacity region. When an extraordinary wave is launched from the inside, the cyclotron and hybrid resonances may be reached for optimal launch angles. In a dense plasma with $\omega_{pe}^2/\omega^2 > 2$, however, there is an opacity region in the center of the plasma column which completely shields the central part of the plasma column and, naturally, "shadows" the part of the hybrid resonance curve near the equator. If the cyclotron resonance condition is not satisfied within the plasma cross section (Fig. 2.7), then the hybrid resonance is always separated from the vacuum by an opacity region.

We now return to the plane layer model and consider the propagation of waves with $N_z^2 \neq 0$, assuming for simplicity that $N_y = 0$. This requires an analysis of the complete dispersion equation (2.25). As can be seen from this equation, the location of the lower hybrid resonance $N_x^2 \rightarrow \infty$ is independent of N_z and is determined, as before, by the condition $\epsilon = 0$, which corresponds to a resonance with perpendicular propagation [see Eq. (2.33)]. This is related to the fact that during wave propagation in a nonuniform plasma layer, N_z always equals its boundary value, so that $N = \sqrt{N_x^2 + N_z^2}$ can go to infinity only as $N_x \rightarrow \infty$ when $\theta \rightarrow \rho/2$.

The cutoff points for the waves, $N_x^2 = 0$ for $N_z^2 \neq 0$, are determined from the condition that the free term of the biquadratic equation (2.25) goes to zero ($A_0 = 0$). This leads to the following equations:

$$\omega_{pe}^2/\omega^2 = 1; \tag{2.43}$$

$$\omega_{pe}^2/[\omega(\omega - |\omega_{Be}|)] = 1 - N_z^2, \tag{2.44}$$

$$\omega_{pe}^2/[\omega(\omega + |\omega_{Be}|)] = 1 - N_z^2. \tag{2.45}$$

From general considerations it is clear that, besides the cutoffs and resonance points, the points where the roots of Eq. (2.25) overlap are distinctive. Near these points the different types of waves in the approximation of geometric optics (ordinary and extraordinary) are coupled. This effect is usually called linear conversion. When $N_z \neq 0$, it can occur under typical tokamak conditions [39, 40]. The situations encountered experimentally are illustrated schematically in Figs. 2.8 and 2.9. Evidently, for small N_z^2 the picture differs little from that of normal wave incidence shown in Figs. 2.4 and 2.5. The main difference is that the cutoff points determined by Eqs. (2.44) and (2.45) are shifted toward the plasma edge. The locations of the cutoffs for the ordinary wave remain unchanged, in accordance with Eq. (2.43). A noticeable difference develops when the cutoffs given by Eqs. (2.43) and (2.45) approach one another (as $N_z^2 \rightarrow |\omega_{Be}|/(\omega + |\omega_{Be}|)|_{n=0}$) (see Fig. 2.8). In this case the opacity region practically disappears for one of the waves launched from the inner edge of the torus (from $x = a$). Then the wave can reach the point x_c, be converted there into another mode, and propagate toward the hybrid resonance. This effect makes it possible in principle to deliver wave energy to the central region of the plasma from the outer edge of the torus when $\omega_{pe}^2(0)/\omega^2 > 1$. In practice, however, it rather difficult to use because of the poloidal magnetic structure and two-dimensional character of the plasma inhomogeneity and because of the relatively narrow range of launch angles over which a strong effect can be expected.

As N_z^2 is increased further, an opacity region again develops. Figure 2.9 shows the case $N_z^2 \gg 1$, where the dispersion of the wave roughly obeys Eq. (2.33). The solution of this equation can be written in the form

$$N_x^2 = -N_z^2 \, \eta/\epsilon. \tag{2.46}$$

These waves can evidently propagate in plasma regions where $\eta/\epsilon < 0$. There are two such regions in a high-density tokamak. One of them corresponds to the conditions $\omega < |\omega_{Be}|$, $\omega < \omega_{pe}$, and the other to the condition $\omega_{pe} < \omega < \omega_{uh}$. These waves are not of interest for plasma heating at $\omega \sim |\omega_{Be}|$, since they cannot propagate in vacuum ($N_x^2 < 0$) and they cannot be excited in practice by antenna structures located on the plasma edge because of the deep opacity barrier.

The question of the field polarization on the plasma edge when $\omega_{pe}^2/\omega^2 \to 0$ is extremely important for exciting this type of wave. For normal incidence on a layer with $N_z = N_y = 0$, the answer is fairly simple: the field of the ordinary wave, as noted above, is directed along the external magnetic field $\mathbf{B}_0$, and the field of the extraordinary wave is perpendicular to it (along the Y axis). When $N_z \neq 0$, however, the answer is considerably more complicated. Then the system of Eqs. (2.24) must be analyzed. Given that $N_x^2 \to 1 - N_z^2$ when $\omega_{pe}^2/\omega^2 \to 0$, from the first of Eqs. (2.24) we obtain

$$E_x/E_z = -N_z/N_x, \tag{2.47}$$

which is the same as the condition for a transverse wave in a vacuum, div $\mathbf{E} = 0$. This equation is valid for both extraordinary and ordinary waves and, therefore, the main difference between the wave types involves the field component E_y. In order to determine E_y one can use the second of Eqs. (2.24), into which it is necessary to substitute N_x^2 calculated with an accuracy corresponding to terms of first order smallness in the parameter ω_{pe}^2/ω^2:

$$\frac{E_y}{E_z} = -\frac{2i\,\omega\,N_z}{\omega_{Be}\,(1-N_z^2)\,N_x} \left(1 \mp \sqrt{1 + \frac{4N_z^2\,\omega^2}{(1-N_z^2)^2\,\omega_{Be}^2}}\right)^{-1}, \tag{2.48}$$

where the minus (upper) sign refers to the extraordinary wave, and the plus (lower) sign, to the ordinary wave.

It is now easy to see how the wave will be polarized when $N_y \neq 0$. This requires a suitable rotation of the coordinate system about the Z axis. Equations (2.47) and (2.48) refer to the case in which the X axis is directed along $\mathbf{k}_\perp$. Rotating this coordinate system so as to obtain the specified projection of $\mathbf{k}_\perp$ along the Y axis [$k_y = (\omega/c)N_y$], we obtain the situation of interest to us. The field components E_x and E_y transform under such a rotation according to well-known formulas.

Studies of wave propagation in cold plasmas using the approximation of geometric optics can, therefore, clarify the conditions for excitation of waves, for their propagation to the plasma center, and for the accessibility of resonances. This approximation, however, does not help in understanding the mechanisms of wave absorption, which to a great extent are associated with thermal motion of the plasma particles.

2.5. ROLE OF THE THERMAL MOTION OF PLASMA PARTICLES (k ∥ B₀)

In Section 2.2 an expression was obtained for the dielectric tensor in a cold plasma. We now examine qualitatively the effects of the thermal motion of the plasma particles. We shall assume initially that a plane wave is propagating in a uniform plasma along an external magnetic field. As in Section 2.2, the direction of the particle velocity induced by a wave field of the form $E \exp[i(k_z z - \omega t)]$ must be calculated. In the linear approximation it is necessary to substitute the position of the particle along its unperturbed trajectory, $z = z_0 + v_z t$, in the exponent, so that the phase of the field acting on a particle will be given by

$$k_z z - \omega t = k_z z_0 - (\omega - k_z v_z)t.$$

In other words, the wave field will act on the particle at a frequency $\tilde{\omega} = \omega - k_z v_z$ that is Doppler shifted relative to ω. Taking this into account, by analogy with Eq. (2.13) we can immediately write

$$u_z = \frac{ie}{m\tilde{\omega}} E_z \exp[i(k_z z - \omega t)]. \tag{2.49}$$

Unlike the cold plasma case, however, the current induced by the wave is not only related to the perturbation in the particle velocity but also to the perturbation in the particle density caused by the wave. For the partial current δj_z, owing to particles with a given velocity v_z, we can write

$$\delta j_z = e[u_z \delta n_0(v_z) + v_z \delta n_1(v_z)], \tag{2.50}$$

where $\delta n_0(v_z)$ is the density of particles with velocity v_z and $\delta n_1(v_z)$ is the density perturbation owing to the wave. For calculating δn_1 one can use the conservation equation for the number of particles

$$\frac{\partial \delta n_1}{\partial t} + \frac{\partial}{\partial z}[u_z \delta n_0 + v_z \delta n_1] = 0,$$

which with Eq. (2.50) gives

$$\delta n_1 = k_z j_z / (e\omega).$$

The resulting expression for δj_z has the form

$$\delta j_z = \frac{ie^2 \delta n_0 \omega}{(\omega - k_z v_z)^2} E_z \exp[i(k_z z - \omega t]. \tag{2.51}$$

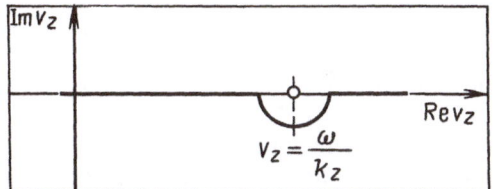

Fig. 2.10. The Landau path rule for bypassing the poles in the integrands.

The total current density in the plasma is obtained by summing the contribution of particles with different v_z. Setting $\delta n_0 = f(v_z)\,dv_z$, where $f(v_z)$ is the longitudinal velocity distribution of the particles, we find

$$\sigma_{zz} = \frac{i\omega_p^2 \omega}{4\pi n_0} \int \frac{f(v_z)\,dv_z}{(\omega - k_z v_z)^2} \, . \tag{2.52}$$

Here the sign for summation over the different types of particles has been left out for brevity. Using the relation $(\omega - k_z v_z)^{-2} = (1/\omega)(d/dv_z)v_z/(\omega - k_z v_z)$, integrating Eq. (2.52) by parts, and using Eq. (2.10), we obtain an expression for the longitudinal component of the dielectric tensor:

$$\epsilon_{zz} = 1 + \frac{\omega_p^2}{\omega n_0} \int \frac{f'(v_z)\,v_z\,dv_z}{\omega - k_z v_z} \, , \tag{2.53}$$

where $f'(v_z) = df(v_z)/dv_z$.

Equation (2.53) is also easily derived by solving the kinetic equation (2.5), which for the case under consideration, $k_x = k_y = 0$, $E_x = E_y = 0$, and $E_z \sim \exp[i(k_z z - \omega t)]$, takes the following form for a Maxwellian unperturbed distribution function $f_0(\mathbf{v})$:

$$-i(\omega - k_z v_z)f_1 = -\frac{e_1}{m}\frac{\partial f_0}{\partial v_z} E_z \, . \tag{2.54}$$

In deriving this equation we have neglected collisions and used the fact that for a Maxwellian distribution function the magnetic field of the wave has no effect on the distribution function (because the vectors $\mathbf{v} \cdot \mathbf{B}$ and $\partial f_0/\partial \mathbf{v} = -(m/T)\mathbf{v}f_0$ are perpendicular). In addition, here the term in Eq. (2.5) containing the unperturbed magnetic field $\mathbf{B}_0$ is also zero. This happens because the field E_z does not perturb the distribution of particle velocities perpendicular to $\mathbf{B}_0$, so that $\partial f_1/\partial \mathbf{v}_\perp = -(m/T)\mathbf{v}_\perp f_1$.

Determining f_1 from Eq. (2.54) and using Eqs. (2.4) and (2.10), it is easy to find a relation of the form (2.53) with $f(v_z) = \int f_0(\mathbf{v})\,d\mathbf{v}_\perp$.

It is evident that for real ω the expressions under the integrals in Eqs. (2.52) and (2.53) have a singularity associated with the presence of resonant particles moving at exactly the phase velocity of the wave. Clearly, the wave field acts on these particles with a constant phase, which is the reason for the resonance. No difficulty arises in calculating the integrals (2.52) and (2.53) if, as before, we use the idea of adiabatically switching on the field and set Im $\omega > 0$. Formally this is equivalent to bypassing the singularities in the complex v_z plane along the contour shown in Fig. 2.10 (the Landau path [29, 32]). The integral (2.53) along the semicircle around the singularity is equal to half the residue at the singularity, and the rest of the integral over real v_z should be taken in the principal value sense. The result is

$$\epsilon_{zz} = 1 + \frac{\omega_p^2}{\omega n_0}\left[\oint \frac{v_z f'(v_z)dv_z}{\omega - k_z v_z} - \frac{i\pi\omega}{k_z^2}f'\left(\frac{\omega}{k_z}\right)\right]. \tag{2.55}$$

The presence of resonant particles, therefore, leads to the appearance of an imaginary part ϵ_{zz}'' in the longitudinal component of the dielectric tensor which corresponds to absorption of the wave energy when $f'(\omega/k_z) < 0$. The rate p of absorption is given in accordance with Eq. (2.22) by

$$p = -\frac{\omega_p^2\omega}{8n_0 k_z^2}f'\left(\frac{\omega}{k_z}\right)|E_z|^2. \tag{2.56}$$

We now analyze Eq. (2.55) for a Maxwellian velocity distribution $f(v_z) = (n_0/\sqrt{\pi}v_T)\exp[-v^2/v_T^2]$, where $v_T = \sqrt{2T/m}$. When $\omega \gg k_z v_T$, the main contribution to the integral in Eq. (2.55) is from the range of v_z over which the condition $\omega \gg k_z v_z$ is satisfied and the denominator under the integral can be expanded in a Taylor series. Integrating the resulting expression term by term, we find

$$\epsilon_{zz} = \eta - \frac{\omega_p^2}{\omega^2}\left[\frac{3}{2}\frac{k_z^2 v_T^2}{\omega^2} + \ldots - i\cdot 2\sqrt{\pi}\frac{\omega^3}{k_z^3 v_T^3}\exp\left(-\frac{\omega^2}{k_z^2 v_T^2}\right)\right], \tag{2.57}$$

where η is ϵ_{zz} in the cold plasma approximation.

In the opposite limiting case of $\omega \ll k_z v_z$, the integral (2.55) is easily evaluated after transforming to the new variable of integration $v = (1/v_T)(v_z - \omega/k_z)$:

$$\frac{1}{n_0}\oint \frac{v_z f'(v_z)dv_z}{\omega - k_z v_z} = \frac{2}{\sqrt{\pi}k_z v_T}\int_{-\infty}^{\infty}\left(v + \frac{\omega}{k_z v_T}\right)^2$$

$$\times \exp\left[-\left(v + \frac{\omega}{k_z v_T}\right)^2\right]\frac{dv}{v}.$$

Expanding the expression under the integral in terms of the small parameter $\omega/k_z v_T$ and carrying out the integration, we obtain the following for the major terms in Eq. (2.55):

$$\epsilon_{zz} = 1 + \frac{2\omega_p^2}{k_z^2 v_T^2}\left(1 + i\sqrt{\pi}\,\frac{\omega}{k_z v_T}\right). \tag{2.58}$$

The cold plasma approximation which we examined previously is, therefore, valid when

$$k_z^2 v_T^2 / \omega^2 \ll 1, \tag{2.59}$$

which ensures that the real and imaginary dispersion corrections to ϵ_{zz} are small.

Although collisionless absorption of wave energy in the presence of resonant particles that are "freely" accelerated by the variable field is not especially surprising when the thermal motion of the particles is taken into account, the way ϵ_{zz}'' depends on the form of the distribution function does require some clarification. In order to examine this question in more detail, we shall calculate the average work q_e done by the field on an isolated electron per unit time:

$$q_e = \langle e\,(v_z + u_z)\,E_z\,(t,\,z^{(0)} + z^{(1)})\rangle,$$

where $z^{(1)}$ is the correction to the unperturbed particle motion $z^{(0)} = z_0 + v_z t$, which is assumed small, so that we can write

$$q_e = e v_z \left\langle \frac{\partial E_z\,(t,\,z^{(0)})}{\partial z^{(0)}}\,z^{(1)}\right\rangle + e\langle u_z\,E_z\,(t,\,z^{(0)})\rangle; \tag{2.60}$$

$z^{(1)}$ can be obtained by integrating the equation of motion in the linear approximation (2.49):

$$z_1 = -\frac{e}{m}\,\mathrm{Re}\,\frac{E_z\,\exp\,[i\,(k_z\,z^{(0)} - \omega't) + \omega''t]}{[(\omega' - k_z v_z) + i\omega'']^2},$$

where the fact that the frequency ω has a small imaginary part corresponding to adiabatic switching on of the field is explicitly taken into account. For particles close to a resonance, in Eq. (2.60) it is possible to retain only the first term, which contains the square of the resonant denominator. Then, dropping the rapidly oscillating terms, we obtain

$$q_e \simeq \frac{e^2 k_z v_z}{m}|E_z|^2\,\frac{\omega''\,(\omega' - k_z v_z)}{[(\omega' - k_z v_z)^2 + \omega''^2]^2}. \tag{2.61}$$

After calculating the energy absorbed per unit time by all the electrons in a unit volume of plasma using $p = \int q_e f(v_z)dv_z$, we obtain an expression identical to Eq. (2.56).

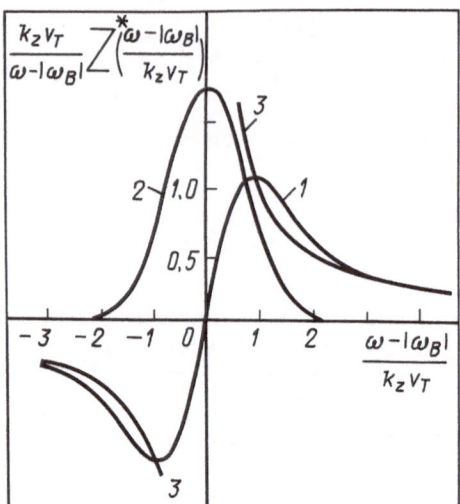

Fig. 2.11. The plasma dispersion function: 1) $\mathrm{Re}\,\dfrac{k_z v_T}{\omega - |\omega_B|}\,Z\!\left(\dfrac{\omega - |\omega_B|}{k_z v_T}\right)$;

2) $-\mathrm{Im}\,\dfrac{k_z v_T}{\omega - |\omega_B|}\,Z\!\left(\dfrac{\omega - |\bar{\omega}_B|}{k_z v_T}\right)$; 3) the function $k_z v_T/(\omega - |\omega_B|)$.

It is evident from Eq. (2.61) that particles moving with a velocity greater than the phase velocity of the wave ($v_z > \omega'/k_z$) give energy to the wave, while particles with a velocity $v_z < \omega'/k_z$ gain energy from the wave. Hence it is clear that the sum, as reflected in Eq. (2.56), depends on the relationship between these two groups of particles, i.e., on the slope of the distribution function at the point $v_z = \omega/k_z$.

Although the sign of q_e depends on the unperturbed velocity v_z, the derivation of Eq. (2.61) shows that the perturbation of the particle positions, $z^{(1)}$, predominates in their energy balance, i.e., spatial clustering of the particles is important. The same conclusion may be reached by analyzing the derivation of Eq. (2.52). In fact, as can be seen from Eq. (2.51), a pole of second order in the integral of Eq. (2.52) is caused by the perturbation in the plasma density, δn_1.

We now consider the effect of the thermal motion on the transverse components of the current density. Since when $\mathbf{k} \parallel \mathbf{B}_0$ motion in a plane perpendicular to $\mathbf{B}_0$ does not perturb the plasma density [div $\mathbf{v}_\perp n = 0$], it is possible to use Eq. (2.17) directly, taking account only of the Doppler frequency shift.* As a result, summing the contribution of the particles with different velocities v_z and using Eq. (2.10), we can write

---

*Strictly speaking, for $v_z \neq 0$ it would also be necessary to take the velocity perturbation caused by the magnetic field of the wave into account. As already noted, however, for a Maxwellian unperturbed velocity distribution, the magnetic field of the wave has no effect on the perturbation in the distribution function or in the plasma current $\mathbf{j}$.

$$\epsilon_{xx} = 1 + \frac{4\pi i \sigma_{xx}}{\omega} = 1 - \frac{\omega_{pe}^2}{2\omega}\left[\frac{1}{\omega - \omega_{Be}} Z\left(\frac{\omega - \omega_{Be}}{k_z v_{Te}}\right)\right.$$

$$\left. + \frac{1}{\omega + \omega_{Be}} Z\left(\frac{\omega + \omega_{Be}}{k_z v_{Te}}\right)\right], \tag{2.62}$$

where, as an example, only the electron part of ϵ_{xx} has been written down and the function $Z(\zeta)$, known as the plasma dispersion function [36], is defined as

$$Z(\zeta) = \frac{v_T \zeta}{n_0} \int \frac{f(v_z)\, dv_z}{v_T \zeta - v_z}. \tag{2.63}$$

This expression is analogous to the integral in Eq. (2.53), so we can write

$$Z(\zeta) = 1 + 1/2\zeta^2 + \dots - i\sqrt{\pi}\,\zeta \exp(-\zeta^2), \quad \zeta^2 \gg 1, \tag{2.64}$$

$$Z(\zeta) = 2\zeta^2 - i\sqrt{\pi}\,\zeta, \quad \zeta^2 \ll 1. \tag{2.65}$$

For weakly slowed-down waves $k_z^2 v_T^2/\omega^2 \ll 1$, therefore, transverse spatial dispersion leads primarily to "smearing out" of the cyclotron resonances and to finite absorption of the electromagnetic wave energy near them. In fact, far from the resonances [for $k_z^2 v_T^2/(\omega - |\omega_B|)^2 \ll 1$], the functions $Z((\omega - |\omega_B|)/k_z v_T)$ and $Z((\omega + |\omega_B|)/k_z v_T)$ differ little from unity (Fig. 2.11), so that Eq. (2.62) transforms to the corresponding expression for a cold plasma (2.19), which grows resonantly as the frequency approaches the cyclotron frequency. For $|\omega - |\omega_B|| \sim k_z v_T$ this growth ceases and a significant imaginary part develops in ϵ_{xx}. For $\omega = |\omega_B|$ the real part of the contribution to ϵ_{xx} from the resonance term goes to zero so that

$$\epsilon_{xx} = 1 + \frac{i\sqrt{\pi}}{2}\frac{\omega_{pe}^2}{\omega k_z v_{Te}} - \frac{\omega_{pe}^2}{4\omega^2}, \quad \omega = |\omega_{Be}|. \tag{2.66}$$

ϵ_{xx} clearly remains finite at the resonance.

2.6. TRANSVERSE SPATIAL DISPERSION (k ⊥ B₀)

We now discuss qualitatively the effect of the thermal motion of particles in the plane perpendicular to B_0, assuming for simplicity that $k_z = 0$ and $k_y = 0$ and considering only the longitudinal component (ϵ_{xx}) of the dielectric tensor. As in Sections 2.2 and 2.5, we shall calculate the part of the velocity of the charges u_z caused by the wave field, beginning with the equation of motion

$$\frac{du_z}{dt} = \frac{e}{m}E_z \exp[i(k_x x - \omega t)]. \tag{2.67}$$

In the absence of a wave, a particle rotates about a cyclotron orbit so that its coordinate x varies as

$$x = x_0 + \rho_B \sin (\omega_B t + \varphi), \tag{2.68}$$

where x_0 is the position of the center of the cyclotron orbit, $\rho_B = v_\perp/\omega_B$ is the cyclotron radius, and φ is the phase of the rotation. Substituting Eq. (2.68) into Eq. (2.67) and using the formula

$$\exp (i \lambda \sin \psi) = \sum_{s=-\infty}^{\infty} J_s (\lambda) \exp (is\psi), \tag{2.69}$$

we obtain

$$\frac{du_z}{dt} = \frac{e}{m} E_z \exp (i k_x x_0) \sum_s J_s (k_x \rho_B) \exp [-i(\omega - s\omega_B) + i\varphi s], \tag{2.70}$$

where $J_s(k_x \rho_B)$ is a Bessel function.

Evidently, the force acting on a particle moving in a cyclotron orbit with $k_\perp \neq 0$ contains a whole spectrum of combination frequencies which differ from the wave frequency by an integral number of harmonics of the cyclotron frequency. Integrating term by term in the sum of Eq. (2.67), we find

$$u_z = \frac{ie}{m} E_z \exp (i k_x x_0) \sum_s \frac{J_s (k_x \rho_B)}{\omega - s\omega_B} \exp [-i (\omega - s\omega_B) t + is\varphi]. \tag{2.70a}$$

To compute the current at a given fixed point x at a given time we must take into account only those particles which have been at the point x prior to time t. Expressing x_0 in terms of x through Eq. (2.68) and again using the expansion (2.69), we write the velocity of the particles passing through x at time t in the form

$$u_z = \frac{ie}{m} E_z \exp [i (k_x x - \omega t)] \sum_{s, l} \frac{J_s (k_x \rho_B) J_l (k_x \rho_B)}{\omega - s\omega_B} \exp [i (s - l)(\omega_B t + \varphi)].$$

This expression is a generalization of Eq. (2.13) to the case $k_x \rho_B \neq 0$. Proceeding as in Section 2.5, but noting that there are no density perturbations, we find the current j_z excited in the plasma by the wave:

$$j_z = \frac{i \omega_p^2}{4\pi n_0} E_z \exp [i (k_x x - \omega t)] \sum_{s, l} \int_0^{2\pi} \frac{\exp [i (s-l)(\omega_B t + \varphi)]}{\omega - s\omega_B} d\varphi$$

$$\times \int_0^\infty f(v_\perp) J_s \left(\frac{k_x v_\perp}{\omega_B} \right) J_l \left(\frac{k_x v_\perp}{\omega_B} \right) v_\perp dv_\perp,$$

where $f(v_\perp) = \int f_0(\mathbf{v})\, dv_z$.

On integrating with respect to φ the terms in this sum with $l \neq s$ go to zero, so that the current has the usual sinusoidal time dependence and

$$\epsilon_{zz} = 1 + \frac{4\pi i \sigma_{zz}}{\omega} = 1 - \frac{\omega_p^2}{\omega n_0} \sum_s \int \frac{J_s^2\left(\frac{k_x v_\perp}{\omega_B}\right) f(v_\perp)\, v_\perp\, dv_\perp}{\omega - s\omega_B} . \tag{2.71}$$

For a Maxwellian $f(v_\perp)$ the integral in Eq. (2.71) can be calculated [29]. This formula then takes the form

$$\epsilon_{zz} = 1 - \frac{\omega_p^2}{\omega} \sum_s \frac{\exp(-\lambda)\, I_s(\lambda)}{\omega - s\omega_B} , \tag{2.72}$$

where $I_s(\lambda)$ is the modified Bessel function and $\lambda = k_x^2 v_T^2/(2\omega_B^2)$.

In an analogous fashion it is possible to obtain the transverse components of the dielectric tensor ϵ_{xx}, $\epsilon_{xy} = -\epsilon_{yx}$, and ϵ_{yy}. Then, however, as in deriving Eq. (2.53), it is necessary to include the perturbation of the particle density by the wave, so the calculations become rather cumbersome. We shall not repeat these calculations here, but give without derivation a formula for ϵ_{xx} which will be needed later:

$$\epsilon_{xx} = 1 - \frac{\omega_p^2}{\omega} \sum_s \frac{s^2 \exp(-\lambda)\, I_s(\lambda)}{\lambda(\omega - s\omega_B)} . \tag{2.73}$$

When the transverse dispersion is taken into account, therefore, terms which are resonant at harmonics of the cyclotron frequency appear in the components of the dielectric tensor. Including the longitudinal spatial dispersion clearly causes "smearing out" of these resonances because of the Doppler effect, so that each resonance term $(\omega - s\omega_B)^{-1}$ in Eq. (2.72) is replaced by an integral of the form $(m/Tn_0 \int v_z^2 f(v_z)\, dv_z/(\omega - s\omega_B - k_z v_z)$, analogous to the integral in Eq. (2.51), while the resonance terms in Eq. (2.73) are replaced by the functions $(1/(\omega - s\omega_B))Z((\omega - s\omega_B)/(k_z v_T))$ [see Eq. (2.63)]. Consequently, at the harmonics of the cyclotron harmonic, anti-Hermitian corrections that cause wave damping appear in the tensor $\epsilon_{\alpha\beta}$. Note that for $k_z \neq 0$ and $k_x \neq 0$ the components $\epsilon_{xz} = \epsilon_{zx}$ and $\epsilon_{yz} = -\epsilon_{zy}$ are also nonzero.

The parameter that determines the role of transverse spatial dispersion is $\lambda = k_\perp^2 v_T^2/(2\omega_B^2)$. When $\lambda \to 0$, all the components of $\epsilon_{\alpha\beta}$ transform to expressions of the type (2.55) and (2.62). In particular, Eq. (2.72) then transforms to Eq. (2.16). For weakly slowed-down waves with $N_x \sim 1$ at high frequencies $\omega \sim |\omega_{Be}|$, the transverse dispersion is usually weak in a nonrelativistic plasma and is generally taken into account through perturbation theory. When the conditions $k_x^2 v_T^2/\omega_B^2 \ll 1$, $k_z v_T^2/\omega^2 \ll 1$, $((\omega - |\omega_B|)/(k_z v_T))^2 \gg 1$, and $((\omega - 2|\omega_B|)/(k_z v_T))^2 \gg 1$ are satisfied, Eqs. (2.72) and (2.73) give

$$\left.\begin{array}{l}
\epsilon_{zz} = \eta - \dfrac{\omega_p^2}{2\omega^2}\left(\dfrac{3k_z^2 v_T^2}{\omega^2} + \dfrac{k_x^2 v_T^2}{\omega^2 - \omega_B^2}\right) ; \\[4mm]
\epsilon_{xx} = \epsilon - \dfrac{\omega_p^2}{2(\omega^2 - \omega_B^2)}\left[\dfrac{(\omega^2 + 3\omega_B^2)k_z^2 v_T^2}{(\omega^2 - \omega_B^2)^2} + \dfrac{3k_x^2 v_T^2}{\omega^2 - 4\omega_B^2}\right].
\end{array}\right\} \quad (2.74)$$

In this approximation the other components have an analogous structure [29]:

$$\left.\begin{array}{l}
\epsilon_{yy} = \epsilon - \dfrac{\omega_p^2}{2(\omega^2 - \omega_B^2)}\left[\dfrac{(\omega^2 + 3\omega_B^2)k_z^2 v_T^2}{(\omega^2 - \omega_B^2)^2} + \dfrac{(\omega^2 + 8\omega_B^2)k_x^2 v_T^2}{\omega^2(\omega^2 - 4\omega_B^2)}\right], \\[4mm]
i\epsilon_{xy} = -g + \dfrac{\omega_p^2}{2\omega(\omega^2 - \omega_B^2)}\left[\dfrac{(3\omega^2 + \omega_B^2)k_z^2 v_T^2}{(\omega^2 - \omega_B^2)^2} + \dfrac{6k_x^2 v_T^2}{\omega^2 - 4\omega_B^2}\right], \\[4mm]
\epsilon_{xz} = -\dfrac{\omega_p^2 k_x k_z v_T^2}{(\omega^2 - \omega_B^2)^2}, \quad \epsilon_{yz} = \dfrac{i}{2}\dfrac{\omega_p^2 \omega_B(3\omega^2 - \omega_B^2)k_x k_z v_T^2}{\omega^3(\omega^2 - \omega_B^2)^2},
\end{array}\right\} \quad (2.75)$$

where ϵ and g are the corresponding quantities in the cold plasma approximation. In Eqs. (2.74) and (2.75) small imaginary contributions proportional to $\exp[-(\omega/k_z v_T)^2]$ and $(k_x^2 v_T^2/\omega_B^2)^{s-1}\exp[-(\omega - s\omega_B)/k_z v_T)^2]$ with $s \geq 1$ have been dropped. For $k_z = 0$ these imaginary contributions are exactly equal to zero in the nonrelativistic approximation being considered here; hence, we conclude that waves propagating perpendicular to the magnetic field are not damped. This conclusion is valid only in the nonrelativistic approximation. It is physically evident that, in addition to the Doppler effect, the relativistic dependence of the cyclotron frequency on the particle velocity can also lead to broadening of the resonances of moving particles. In the weakly relativistic case, $v_T^2/c^2 \ll 1$, we can write $\omega_B(v) = \omega_{B0}(1 - v^2/2c^2)$ and the resonance terms of the form $(\omega - |\omega_B|)^{-1}$ in Eq. (2.62) must be replaced by

$$\frac{1}{\omega - |\omega_B|}\tilde{Z}\left[\frac{(\omega - |\omega_{B0}|)c^2}{|\omega_{B0}|v_T^2}\right] = \frac{1}{n_0}\int\frac{f_0(v)dv}{\omega - |\omega_{B0}|(1 - v^2/(2c^2))},$$

where, by analogy with Eq. (2.63), the function $\tilde{Z}$ has been introduced to describe the relativistic broadening of the cyclotron resonance. To within the accuracy of interest to us, small quantities of order v^2/c^2 can be neglected everywhere except in the resonance denominators, so that the nonrelativistic Maxwellian distribution function can be used for $f_0(v)$. Then, transforming to spherical coordinates in velocity space in the integral for $\tilde{Z}$, we obtain

$$\frac{1}{\omega - |\omega_{B_0}|} \tilde{Z}\left[\frac{(\omega - |\omega_{B_0}|)c^2}{|\omega_{B_0}|v_T^2}\right] = \frac{4}{\sqrt{\pi}\,v_T^3} \int \frac{\exp(-v^2/v_T^2)\,v^2\,dv}{\omega - |\omega_{B_0}| + (|\omega_{B_0}|/(2c^2))\,v^2} . \qquad (2.76)$$

For $\omega < |\omega_B|$ the expression under the integral in Eq. (2.76) has a pole. Here the same integration path rule employed for computing Landau damping can be used. As a result, the imaginary part of the integral (2.76) reduces to half of the residue at the pole:

$$\mathrm{Im}\,\frac{1}{\omega - |\omega_{B_0}|} \tilde{Z} = \frac{4\sqrt{2\pi}\,c^3}{v_T^3(|\omega_{B_0}|)^{3/2}} \sqrt{|\omega_{B_0}| - \omega} \times \exp\left[-\frac{2c^2\,(|\omega_{B_0}| - \omega)}{|\omega_{B_0}|v_T^2}\right] . \qquad (2.77)$$

Structurally similar expressions would follow from Eq. (2.71) in the neighborhood of the resonance at an arbitrary harmonic of the cyclotron frequency ($\omega \to |s\omega_{B_0}|$) with the sole difference that for the lowest order in v_T^2/c^2 there would be an additional factor of the form

$$\frac{1}{2}\int_0^\pi J_s^2\left(\frac{k_x c}{|\omega_{B_0}|}\sqrt{\frac{2(|s\,\omega_{B_0}| - \omega)}{|s\,\omega_{B_0}|}}\sin\vartheta\right)\sin\vartheta\,d\vartheta .$$

It is clear from Eq. (2.77) that relativity leads to substantial changes in the analytic properties of the components of $\epsilon_{\alpha\beta}$ as functions of the frequency: instead of a pole at $\omega = |\omega_B|$ they have a branch point. Wave damping owing to a resonance at harmonics of the cyclotron frequency can only occur when $\omega < |s\omega_{B_0}|$ ($s = 1, 2$), since for $\omega > |s\omega_{B_0}|$ there is no pole under an integral of the form (2.77) and the function $\tilde{Z}$ is real. This damping behavior is fully understandable from a physical standpoint: the relativistic corrections increase the particle mass so that the cyclotron frequency is greatest for particles at rest and the possibility of a resonance with the wave on the part of some fraction of the particles merely requires that $\omega < |\omega_{B_0}|$.

When $N_z = k_z c/\omega \neq 0$, the condition for resonance of a particle with the wave obviously has the form

$$\omega(1 - N_z v_z/c) - |s\omega_{B0}| [1 - (v_z^2 + v_\perp^2)/(2c^2)] = 0 \qquad (2.78)$$

and it may be concluded that relativistic effects are unimportant when the inequality

$$N_z \gg v_T/c \qquad (2.79)$$

is satisfied. A rigorous analysis [41] shows, however, that the limiting transition to the nonrelativistic case is nontrivial and, strictly speaking, the criterion (2.79) is not sufficient. The character of the difficulties which may then arise can be illustrated as follows. It is easily shown that Eq. (2.78) can be satisfied only if $\omega^2 \leq (s\omega_{B0})^2 + N_z^2$. (For simplicity it is assumed that $N_z^2 \ll 1$.) When $\omega^2 > (s\omega_{B0})^2 + N_z^2$, a resonance is impossible for any particle velocity, so there is no wave damping whatever. Hence, when $\omega^2 = (s\omega_{B0})^2 + N_z^2$, the tensor $\epsilon_{\alpha\beta}$ has a branch point near which it has completely different analytic properties than in the nonrelativistic approximation. When $N_z^2 \gg v_T^2/c^2$, however, the branch point of $\epsilon_{\alpha\beta}(\omega)$ corresponds to the far "wing" of the nonrelativistic absorption profile where the main contribution to $\epsilon_{\alpha\beta}$ is from particles with $v_z < cN_z$, for which the relativistic correction is of little importance. Here the thermal corrections are small and it is possible to use the cold plasma approximation. The above-mentioned difficulty, therefore, only arises when an accurate accounting of the thermal corrections to $\epsilon_{\alpha\beta}$ is necessary.

We conclude with a comment: in calculating the variable components of the particle velocity we have neglected the effect of the magnetic field of the wave. Nevertheless, as noted above, averaging the resulting formulas over a Maxwellian distribution function yields the correct result for the current density induced in the plasma by the wave. This does not at all mean that the velocities of the individual particles were correctly calculated. Furthermore, some of the results may lead to errors. This refers, in particular, to Eq. (2.70a) which implies that particles may be resonantly accelerated along an external magnetic field $\mathbf{B}_0$ when $\omega \to |\omega_B|$. By including a magnetic field $\mathbf{B} = (c/\omega)(\mathbf{k} \times \mathbf{E})$ in the equation of motion for the particles, it is easy to confirm that the resonance term in the electric part of the Lorentz force with eE_z is compensated by the term $(e/c)(\mathbf{v} \times \mathbf{B})_z$ while the term $(e/c)(\mathbf{v} \times \mathbf{B})_x = (e/c)v_xB_y$ leads to a resonant change in the transverse particle velocity when $\omega \to |\omega_B|$. In other words, during cyclotron absorption of an ordinary wave with $\mathbf{E} \parallel \mathbf{B}_0$ and $\mathbf{k} \perp \mathbf{B}_0$, the wave energy is transformed into kinetic energy of the particles associated with their motion perpendicular to $\mathbf{B}_0$. The increase in the longitudinal plasma current at a given point and time, on the other hand, is not due to an increase in the longitudinal particle velocity, but to a spatial redistribution of the particles owing to the perturbing effect of the wave on their cyclotron rotation. Because of this perturbation, at this point there is an excess of particles whose unperturbed velocity is in the direction required to produce the current.

2.7. EFFECT OF SPATIAL DISPERSION ON RESONANCES

We now examine the effect of spatial dispersion on the propagation of plane waves in a uniform plasma, emphasizing resonances. It is clear that Eq. (2.9) is valid for arbitrary spatial dispersion. The system of Eqs. (2.24) and the corresponding dispersion equation (2.25) are valid only for $\epsilon_{\alpha\beta}$ with the special form (2.21). Therefore, Eqs. (2.24) and (2.25) can be used directly in the presence of spatial dispersion only in the case $k_\perp = 0$, when $\epsilon_{\alpha\beta}$ has the form of Eq. (2.21) with ϵ, g, and η replaced by the corresponding expressions from Sections 2.5 and 2.6. Even in the general case, however, the dispersion equation may be written in the form (2.25), with the sole difference that the coefficients A_2 and A_0 include additional terms containing the components ϵ_{yz}, ϵ_{zy}, ϵ_{xz}, ϵ_{zx}, and $\epsilon_{yy} - \epsilon_{xx}$. (The coefficient for N_x^4 will be ϵ_{xx}.)

We first examine space charge waves propagating along the magnetic field, for which, in accordance with Eq. (2.26),

$$\epsilon_{zz} = 0. \tag{2.80}$$

When $\omega_{pe}^2 \gg k_z^2 v_{Te}^2$, using Eq. (2.53) we find the approximation

$$\omega^2 \simeq \omega_{pe}^2 + \frac{3}{2} k_z^2 v_{Te}^2 - 2i\sqrt{\pi} \frac{\omega_{pe}^5}{k_z^3 v_{Te}^3} \exp\left[-\frac{3}{2} - \frac{\omega_{pe}^2}{k_z^2 v_{Te}^2}\right]. \tag{2.81}$$

An important difference from the cold plasma case can be seen: the frequency of the wave now depends on the wave vector, and finite damping of the wave also appears. The damping is weak only when $\omega_{pe}^2/(k_z^2 v_{Te}^2) \gg 1$.

There is yet another important difference. It follows from Section 2.5 that the electron term in ϵ_{zz} is bounded by $2\omega_{pe}^2/(k_z^2 v_{Te}^2)$ when $\omega \to 0$, while for cold ions the ion term goes to negative infinity as ω^{-2}. Therefore, at some frequency these terms may compensate one another and yet another solution of Eq. (2.80) will be possible. In fact, using Eq. (2.58) and taking the ion contribution into account, it is possible to transform Eq. (2.80) to the form

$$1 + \frac{2\omega_{pe}^2}{k_z^2 v_{Te}^2} - \frac{\omega_{pi}^2}{\omega^2} + i \frac{2\sqrt{\pi}\,\omega_{pe}^2\,\omega}{k_z^3 v_{Te}^3} = 0.$$

Since usually $\omega_{pe}^2/(k_z^2 v_{Te}^2) \gg 1$, this equation has the approximate solution

$$\omega^2 = k_z^2 c_s^2 - i\sqrt{\pi}\sqrt{m_e/(2m_i)}\,\omega^2, \tag{2.82}$$

where the fact that $\omega_{pe}^2/\omega_{pi}^2 = m_i/m_e$ was used (for singly charged ions) and $c_s = \sqrt{T_e/m_i}$ is the ion sound speed.

Taking spatial dispersion into account, therefore, leads to the appearance of yet another solution of the dispersion equation (2.25), one that corresponds to ion acoustic waves. This is quite natural since when the dependence of the coefficients in Eq. (2.25) on k_z is taken into account, it becomes transcendental, generally speaking, with an infinite number of complex solutions.

We now examine the effect of spatial dispersion on the propagation of rf waves in the neighborhood of the lower hybrid resonance. Since $k_x \to \infty$ in a cold plasma as this resonance is approached (i.e., as $\epsilon \to 0$), it is natural in the first approximation to try to include the dispersion corrections only in ϵ_{xx} [the coefficient of N_x^4 in Eq. (2.25)] and to retain only the terms which are proportional to k_x^2 in these corrections. We shall do this as an illustration for the case $k_z = 0$. In this approximation one can use Eq. (2.29) with ϵ on the right-hand side replaced by ϵ_{xx} from Eq. (2.74) :

$$(\epsilon - \beta_{\text{eff}} N_x^2) N_x^2 = \epsilon^2 - g^2, \tag{2.83}$$

where $\beta_{\text{eff}} = (3v_{Te}^2/2c^2)\omega_{pe}^2\omega^2/[(\omega^2 - \omega_{Be}^2)(\omega^2 - 4\omega_{Be}^2)]$. As a rule, under tokamak conditions $|\beta_{\text{eff}}| \ll 1$ outside the electron cyclotron resonance region, so that far from hybrid resonance ($\epsilon^2 \gg |\beta_{\text{eff}}|$) Eq. (2.83) has the approximate solutions

$$N_x^2 \simeq (\epsilon^2 - g^2)/\epsilon + 0(\beta_{\text{eff}}) \tag{2.84}$$

and

$$N_x^2 \simeq \epsilon/\beta_{\text{eff}}. \tag{2.85}$$

The first of these clearly describes the extraordinary wave for a cold plasma [Eq. (2.29)] with a minor thermal correction, and the second, which coincides with the solution to Eq. (2.33), describes a longitudinal plasma wave caused by the thermal motion of the particles. When $\epsilon^2 \to 4\beta_{\text{eff}}g^2$, these two solutions merge continuously into one another.

Examining the propagation of these waves in the nonuniform plasma layer described in Section 2.4 in the approximation of geometric optics, we obtain a dispersion curve N_x^2 (Fig. 2.5). It is apparent that when an extraordinary wave is incident on the hybrid resonance layer, it is converted into a plasma wave which propagates into the depth of the plasma. From this it is clear that the collisional wave absorption at the hybrid resonance discussed in Section 2.4 actually describes a linear wave conversion, i.e., the transfer of energy from the fast electromagnetic to a slow plasma wave which degenerates into space charge eigenmodes in the cold plasma limit. Linear wave conversion in the neighborhood of the hybrid resonance has been discussed by many authors in connection with rf heating of plasmas [37, 38]. Various aspects of this phenomenon have been studied over a wide range of conditions: conversion efficiency, field structure, the mechanism for energy dissipation, etc. These details are not so important for the qualitative picture of wave phenomena presented here. We note only that in most cases of practical interest, the conversion efficiency is independent of the relationship among the small parameters of the problem, which characterize the spatial dispersion and wave absorption (see [37]). The energy carried away by the plasma wave in the absence of dissipation is equal to the energy absorbed in the neighborhood of the resonance in a cold plasma. Thus, the conversion efficiency can be evaluated by solving the simpler problem of wave absorption in the cold plasma approximation. In the particular case being examined here, the conversion efficiency and absorption are, as noted in Section 2.4, determined primarily by the opacity barrier.

Equation (2.85) implies that, depending on the sign of β_{eff}, a plasma wave may propagate either into the depth of the plasma in the direction of increasing magnetic field (for $\beta_{\text{eff}} < 0$) or toward the edge of the plasma column (for $\beta_{\text{eff}} > 0$). In both cases, the refractive index of the plasma increases rapidly with distance from the

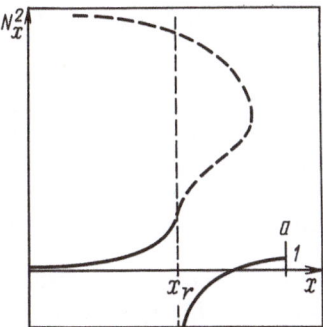

Fig. 2.12. The variation in the square of the perpendicular refractive index of an extraordinary wave (——) and of a longitudinal wave (- - -) in a nonuniform plasma when the thermal motion of the particles is taken into account.

point where $\epsilon(x) = 0$, and the approximation of weak spatial dispersion, used to derive Eqs. (2.74) and (2.83), may no longer be valid. Using Eq. (2.85), the condition for applicability of this approximation, $k_x^2 v_{Te}^2/\omega_{Be}^2 \ll 1$, can be written in the form

$$\frac{\omega^2 |\omega^2 - \omega_{Be}^2||\omega^2 - 4\omega_{Be}^2|}{\omega_{Be}^2 \, \omega_{pe}^2} |\epsilon| \ll 1.$$

Thus, in the high-frequency region $\omega \sim |\omega_{Be}| \sim \omega_{pe}$ Eq. (2.85) is valid only in a small neighborhood of the hybrid resonance with $|\epsilon| \ll 1$. Outside this region the spatial dispersion must be taken into account more accurately. The complete dispersion equation (2.11) and the exact expressions for $\epsilon_{\alpha\beta}$ must be used for a rigorous solution of the problem. The problem can be greatly simplified by noting that the field of the plasma waves is practically longitudinal because of the large refractive index, so that Eq. (2.33) can be used to describe the dispersion of the plasma waves. As noted previously, Eq. (2.85) already satisfies this equation. It is even more valid far from the hybrid resonance, with still larger values of N_x^2.

In accordance with Eq. (2.73) with $N_x = 0$, Eq. (2.33) takes the form

$$1 - \sum_{s=1}^{\infty} \frac{2\omega_{pe}^2 \, s^2 \exp(-\lambda) I_s(\lambda)}{\lambda(\omega^2 - \omega_{Be}^2)} = 0, \tag{2.86}$$

where $\lambda = \omega^2 v_{Te}^2 N_x^2/(2\omega_{Be}^2 c^2)$. When $\beta_{eff} < 0$ (i.e., $|\omega_{Be}| < \omega < 2|\omega_{Be}|$), the wave, as mentioned above, propagates from the conversion point ($\epsilon \approx 0$) in the direction of increasing magnetic field, thereby approaching the cyclotron resonance point. In order for the nonresonant terms in Eq. (2.86) to compensate the resonance term in the neighborhood of the resonance, the factor in it which depends on N_x^2 ($e^{-\lambda} I_1(\lambda)\lambda^{-1}$) must approach zero as $\omega^2 \to \omega_{Be}^2$. This is possible only when $\lambda \to \infty$. Using the asymptotic representation for the function $I_1(\lambda)$ and neglecting the small terms with $s > 1$ in Eq. (2.86), we find

$$\sqrt{2\pi\lambda^3} = 2\omega_{pe}^2/(\omega^2 - \omega_{Be}^2).$$

For $\beta_{eff} < 0$, therefore, the refractive index of a plasma wave increases monotonically as it propagates from the conversion point into the depth of the plasma and goes to infinity at the cyclotron resonance point. Of course, near the resonance it is necessary to take either relativistic effects or Doppler broadening of the resonance into account because of the finite value of N_z. In any case, complete absorption of the wave will occur here.

Equation (2.86) can also be used to track the refractive index of the plasma wave for $\beta_{eff} > 0$ (i.e., for $\omega^2 > 4\omega_{Be}^2$). This is especially easy to do for $\omega^2/(9\omega_{Be}^2 - \omega^2) \gg 1$, when Eq. (2.86) can be written approximately in the form

$$\beta_2 N^4 - \beta_{eff} N^2 + \epsilon = 0,$$

where $\beta_2 \simeq \dfrac{243}{32} \dfrac{\omega_{pe}^2}{9\omega_{Be}^2 - \omega^2} \dfrac{v_{Te}^4}{c^4} > 0$. This shows that the plasma wave formed by conversion initially propagates from the hybrid resonance point in the direction of decreasing plasma density and, in turn, is converted when $\epsilon = \beta_{eff}^2/4\beta_2$ into a slower wave that propagates into the depth of the layer (Fig. 2.12).

The strongly slowed-down waves described by Eq. (2.86) are often referred to as Bernstein modes in the non-Soviet literature. A detailed analysis of their dispersion properties has been given by Akhiezer et al. [29].

In conclusion, we briefly discuss the effect of the thermal motion of the particles on the electron cyclotron resonance. (For a detailed discussion of this topic see [29].) As shown in Sections 2.3 and 2.4, in the cold plasma approximation there is no resonant component of the electric field at all in the wave because of Eq. (2.30). Since ϵ_{xx} and ϵ_{xy} do not go to infinity at the resonance point when the thermal motion is taken into account and $N_z \neq 0$, the quantity $|g/\epsilon|$ will not be exactly equal to unity. The ratio of the "nonresonance" terms in the first approximation of Eq. (2.30) to ϵ_{xx} will also not go to zero, so that Eq. (2.30) will contain corrections $\sim 1/\epsilon_{xx}$. As a result, the wave will have an admixture of resonance polarization. The fraction of this admixture will depend on $|\epsilon_{xx}|$ at the resonance. It is easy to understand that this fraction will be inversely proportional to

$$\left| \frac{\omega_{pe}^2}{\omega(\omega - |\omega_{Be}|)} Z\left(\frac{\omega - |\omega_{Be}|}{k_z v_{Te}}\right) \right| \simeq \frac{\omega_{pe}^2}{\omega k_z v_{Te}} \gg 1.$$

Evidently, the presence of a resonance field component rotating in the direction of the cyclotron rotation of a particle will lead to absorption of wave energy when $|\omega + \omega_{Be}| \lesssim k_z v_{Te}$. Let us first make a rough estimate of the efficiency of this absorption for the extraordinary wave. We shall consider the case of "quasi-perpendicular" propagation $N_z^2 \ll 1$. From intuitive considerations one may conclude that for sufficiently small N_z^2 in the first approximation it is possible to use Eq. (2.29), where the components of the tensor are calculated with the longitudinal spatial dispersion taken into account. This equation and the formulas of Section 2.5 then show that in the resonance region $(\omega + \omega_{Be})^2 < k_z^2 v_{Te}^2$, the imaginary part of the refractive index can be estimated as follows:

$$\operatorname{Im} N_x \sim \frac{\epsilon_{xx} - i\epsilon_{xy}}{N_x} \operatorname{Im} \frac{\epsilon_{xx} + i\epsilon_{xy}}{\epsilon} \sim \frac{\omega\, k_z\, v_{Te}}{\omega_{Be}^2} .$$

This estimate becomes completely obvious if the cold plasma approximations for ϵ_{xx} and ϵ_{xy} are used with the resonance denominator $(\omega + \omega_{Be})^{-1}$ replaced by $-i/(k_z v_{Te})$. In a tokamak the width of the resonance zone ΔR is determined by the condition

$$\frac{d|\omega_{Be}|}{dx}\, \Delta R \simeq \omega\, \frac{\Delta R}{R_0} \sim 2k_z\, v_{Te} . \tag{2.87}$$

Hence, the optical thickness Γ for the extraordinary wave can be written in the form

$$\Gamma \simeq \Delta R \operatorname{Im} k_x = 2\, \frac{\omega R_0}{c}\, \frac{k_z^2\, v_{Te}^2}{\omega_{pe}^2}\, N_x. \tag{2.88}$$

It follows from Section 2.6 that this equation is applicable only for $k_z \gg \omega v_{Te}/c^2$. When $k_z \simeq \omega v_{Te}/c^2$, relativistic broadening of the resonances plays an important role. Nevertheless, Eq. (2.88) can be used to estimate Γ when $k_z \to 0$ if we set $k_z \sim \omega v_{Te}/c^2$.

Equation (2.88) shows that the optical thickness of the resonance layer for the extraordinary wave falls with increasing density. This reflects the fact that in a denser plasma the resonant component of the field is more strongly shielded. We note that Eq. (2.88) is valid for a rather dense plasma in which $\omega_{pe}^2/(\omega k_z v_{Te}) \gg 1$. In a rarefied plasma with $\omega_{pe}^2/(\omega k_z v_{Te}) \ll 1$ the imaginary part of ϵ_{xx} is a small correction, so that $\operatorname{Im} N_x \sim \omega_{pe}^2/(k_z v_{Te}\omega)$ and we obtain

$$\Gamma = (2\omega R_0/c)\, \omega_{pe}^2 / \omega^2 .$$

instead of Eq. (2.88). Here the resonant component of the field is not screened and the weak damping is simply a consequence of the paucity of particles in the plasma. When $\omega_{pe}^2/(\omega k_z v_{Te}) \sim 1$, damping of the extraordinary wave becomes strong $(\operatorname{Im} N_x \sim N_x)$.

In the case quasiperpendicular propagation of the ordinary wave it is possible to use the dispersion relation

$$N_x^2 = \epsilon_{zz} = 1 - \frac{\omega_{pe}^2}{\omega^2} - \frac{\omega_{pe}^2}{\omega(\omega + \omega_{Be})}\, \frac{k_x^2\, v_{Te}^2}{4\omega_{Be}^2}$$

and note that longitudinal spatial dispersion causes broadening of the resonance. In accordance with Section 2.5, a rough estimate of the effect of spatial dispersion can be obtained, as in the previous case, by replacing the resonance denominator $(\omega + \omega_{Be})^{-1}$ in the formula for N_x^2 by $-i/(k_z v_{Te})$. Then, finding $\mathrm{Im}\, N_z$ and using Eq. (2.87), we obtain

$$\Gamma \sim \frac{\omega R_0}{c} \frac{\omega_{pe}^2}{\omega^2} \frac{v_{Te}^2}{4c^2} (1 - \frac{\omega_{pe}^2}{\omega^2})^{1/2}. \tag{2.89}$$

It is interesting that this estimate is correct even for $N_z < v_{Te}/c$, when relativistic effects must be taken into account. This agreement between the relativistic and non-relativistic theories is not by chance and is related to the analytic properties of the components of $\epsilon_{\alpha\beta}$ as functions of the frequency.

Equation (2.89) can be used for $N_z^2 \ll 1$. The damping of the ordinary wave falls off rapidly as N_z^2 increases.

2.8. PROPAGATION OF WAVES WITH $\omega \ll |\omega_{Be}|$. THE LOWER HYBRID RESONANCE

As in our discussion of high frequencies, we begin with the perpendicular propagation of waves $(k_z = 0)$ in the cold plasma approximation. Then, as in Section 2.3, is is possible to use Eqs. (2.29) and (2.34) with the contribution of the ions included in the dielectric tensor. It follows from Sections 2.3 and 2.4 that the physics of wave propagation is determined to a great extent by the location of the cutoffs and hybrid resonances. The latter are determined by the condition

$$\epsilon = 1 - \sum_j \omega_{pj}^2 / (\omega^2 - \omega_{Bj}^2) = 0. \tag{2.90}$$

We now determine the frequencies at which this equality can hold in tokamak experiments. Assume for now that there is only one ion species in the plasma. In the rf region, Eq. (2.90) is satisfied for $\omega = \omega_{uh}$. When $\omega < |\omega_{Be}|$, the electron contribution to ϵ becomes positive, so that at low frequencies Eq. (2.90) can be satisfied only because of the ions. Since $\omega_{pi}^2/\omega_{pe}^2 = Z_i m_e/m_i \ll 1$ (where Z_i is the ionic charge), the ions can be important only at fairly low frequencies $\omega^2 \ll \omega_{Be}^2$, so that we can write approximately

$$\epsilon \simeq 1 + \omega_{pe}^2/\omega_{Be}^2 - \omega_{pi}^2/(\omega^2 - \omega_{Bi}^2). \tag{2.91}$$

In this approximation Eq. (2.90) has one solution:

$$\omega^2 = \omega_{LH}^2 \equiv \omega_{Bi}^2 + \omega_{pi}^2 \omega_{Be}^2/(\omega_{Be}^2 + \omega_{pe}^2). \tag{2.92}$$

The resonance at ω_{LH} is called the lower hybrid resonance. Usually in tokamaks $\omega_{pi}^2(0) \gg \omega_{Bi}^2$, and the lower hybrid resonance frequency satisfies the conditions $\omega_{Be}^2 \gg \omega_{LH}^2 \gg \omega_{Bi}^2$, so that Eqs. (2.91) and (2.92) can be written in the form

$$\epsilon = 1 + \omega_{pe}^2/\omega_{Be}^2 - \omega_{pi}^2/\omega^2 \tag{2.91a}$$

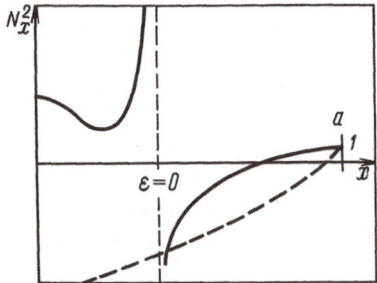

Fig. 2.13. The propagation behavior of ordinary (——) and extraordinary (- - -) waves in the lower hybrid resonance frequency range with $N_z = 0$.

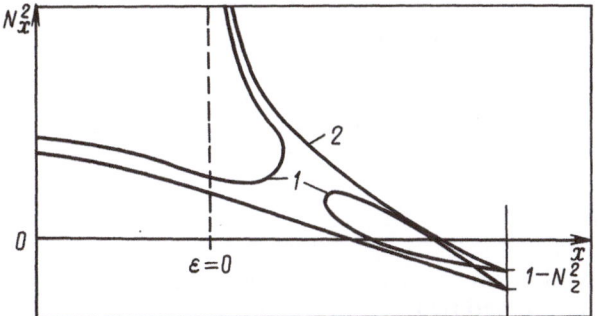

Fig. 2.14. The behavior of the square of the perpendicular refractive index of lower hybrid waves for different amounts of longitudinal slowing-down N_z: 1) $N_z^2 < 1 + (\omega_{pe}^2/\omega_{Be}^2)_{LH}$; 2) $N_z^2 > 1 + (\omega_{pe}^2/\omega_{Be}^2)_{LH}$.

and

$$\omega_{LH}^2 = \omega_{pi}^2 \, \omega_{Be}^2 / (\omega_{Be}^2 + \omega_{pe}^2). \tag{2.92a}$$

Equation (2.91a) is also applicable to plasmas with different ion species if ω_{pi}^2 is taken to mean $\Sigma_j \omega_{pij}^2$.

Since ω is set by the heating source and the density varies over the plasma cross section, the localization of the hybrid resonance (i.e., of the density at which $\omega = \omega_{LH}$) in the plasma is of interest. From Eq. (2.91a) we find

$$(\omega_{pe}^2/\omega_{Be}^2)_{LH} = \omega^2/(|\omega_{Be}|\omega_{Bi} - \omega^2). \tag{2.93}$$

The conditions for the lower hybrid resonance can exist only when the inequality

$$|\omega_{Be}|\omega_{Bi} > \omega^2 \tag{2.94}$$

is satisfied. Since the parameter $\omega_{pe}^2(0)/\omega_{Be}^2$ does not differ greatly from unity in tokamaks, this inequality should not be very strong.

In a plasma with one ion species, therefore, ϵ goes to zero only at the upper and lower hybrid frequencies. The situation changes when different kinds of ions with different cyclotron frequencies exist in the plasma. As an example, consider two ion species with $\omega_{B1} > \omega_{B2}$. In the frequency range $\omega_{B1} > \omega > \omega_{B2}$ the contributions of the different ions to ϵ will differ in sign and there is yet another possible hybrid resonance. To find the frequency of this ion–ion hybrid resonance we note that in this frequency range, the ion contribution is dominant in ϵ, so that the term which equals unity and the electron term can be neglected. Introducing the relative concentration $\zeta = Z_1 n_1/n_e$ of the first ion species (with charge state Z_1) and noting that $\omega_{B1}\omega_{p2}^2/\omega_{B2}\omega_{p1}^2 = (1 - \zeta)/\zeta$, we transform Eq. (2.90) to

$$\epsilon \simeq \left[\frac{1}{\omega^2 - \omega_{B2}^2} + \frac{\zeta \omega_{B1}}{\omega_{B2}(1-\zeta)(\omega^2 - \omega_{B1}^2)} \right] \omega_{p2}^2 = 0. \tag{2.95}$$

This yields the ion–ion hybrid frequency ω_{ii}:

$$\omega_{ii}^2 = \omega_{B1}\omega_{B2} \frac{(1-\zeta)\omega_{B1} + \zeta \omega_{B2}}{(1-\zeta)\omega_{B2} + \zeta \omega_{B1}}. \tag{2.96}$$

Depending on the relative concentration of the ions, the frequency of this resonance varies from ω_{B1} (when $\zeta = 0$) to ω_{B2} (when $\zeta = 1$).

We now consider the wave cutoffs at $N_z = 0$. According to Eqs. (2.29) and (2.34), they are determined by the conditions

$$\epsilon \pm g = 1 - \sum_j \omega_{pj}^2/\omega(\omega \pm \omega_{Bj}) = 0 \tag{2.97}$$

and

$$\eta = 1 - \omega_{pe}^2/\omega^2 = 0. \tag{2.98}$$

For frequencies $\omega^2 \ll \omega_{Be}^2$ the formulas for $\epsilon \pm g$ in a plasma with a single ion species can be transformed to

$$\epsilon \pm g = 1 \mp \omega_{pe}^2/|\omega_{Be}|(\omega \mp \omega_{Bi}). \tag{2.99}$$

The inequality (2.97) is clearly satisfied at the point where

$$\omega_{pe}^2/|\omega_{Be}|(\omega - \omega_{Bi}) = 1. \tag{2.100}$$

Given the locations of the cutoffs and resonance points, it is easy to describe generally how the refractive indices of geometric optics will vary for these waves in the plasma layer discussed in Section 2.4 (Fig. 2.13). It is important to note that here the hybrid resonance region is separated from the plasma edge by a wide opacity region, so that waves excited in the vacuum cannot penetrate into the center of the plasma.

In order to study the propagation of waves with $N_z \neq 0$, we must turn to the complete dispersion equation (2.25) or to the system of equations (2.24). It is convenient to begin analyzing the possible solutions of these equations with the case $N_z^2 \gg 1$, noting explicitly that when $\omega^2 \ll \omega_{Be}^2$ the component η is much greater than all the other components of the dielectric tensor. As can be seen from the third equation of Eqs. (2.24), a large value of $|\eta|$ means that the longitudinal component of the electric field E_z is very small. As a result, Eqs. (2.24) can be reduced approximately to the following form:

$$\left. \begin{array}{l} (\epsilon - N_z^2)E_x + ig\,E_y = 0, \\[2mm] (\epsilon - N_z^2 - N_x^2)E_y - ig\,E_x = 0, \quad E_z \simeq 0. \end{array} \right\} \tag{2.101}$$

The smallness of E_z at low wave frequencies is easy to understand qualitatively: electrons moving freely along the external magnetic field $\mathbf{B_0}$ "short out" the component of the electric field parallel to $\mathbf{B_0}$.

Eliminating E_x from the system (2.101), we obtain the dispersion equation

$$N_x^2 = [(\epsilon - N_z^2)^2 - g^2]/(\epsilon - N_z^2). \tag{2.102}$$

A second solution of Eq. (2.25) for $N_z^2 \gg 1$ can be obtained starting with the idea that a wave with a large refractive index is longitudinal. Then Eq. (2.33a) can be used in the first approximation. A somewhat more accurate result is obtained if the field component E_y, which is zero for longitudinal waves ($\mathbf{E} = -\nabla\varphi$), is neglected in Eqs. (2.24). Then the system (2.24) takes the form

$$\left. \begin{array}{l} (\epsilon - N_z^2)E_x + N_x N_z\, E_z = 0, \\[2mm] (\eta - N_x^2)E_z + N_x N_z\, E_x = 0 \end{array} \right\} \tag{2.103}$$

and then

$$N_x^2 = \eta(\epsilon - N_z^2)/\epsilon. \tag{2.104}$$

The waves described by Eqs. (2.102) and (2.104) are referred to as fast and slow, respectively, in the lower hybrid frequency range $\omega \sim \sqrt{|\omega_{Be}|\omega_{Bi}|}$. Sometimes they are called the H- and E-waves (from their polarization at the edge of the plasma). In the low-frequency range the first of these is also called the fast magnetosonic wave. The slow wave with $N_z^2 \gg 1$ is often called the oblique Langmuir wave. This is because when $\omega_{pe}^2/\omega_{Be}^2 \ll 1$ the solution of Eq. (2.104) can be written approximately as $\omega = \omega_{pe}\cos\theta$, which is analogous to Eq. (2.26).

The behavior of the refractive index of the fast and slow waves in a nonuniform plasma layer with $\omega_{Bi}^2 \ll \omega^2 \ll \omega_{Be}^2$ and $N_z \gg 1$ is shown in Fig. 2.14 (curve 2). The edge of the plasma is opaque to both waves. The cutoff for the fast wave [Eqs. (2.102) and (2.99)] is given by $\omega_{pe}^2/|\omega|\omega_{Be}| = N_z^2 - 1$. The slow wave is cut off at the critical density ($\omega_{pe}^2 = \omega^2$). At the hybrid resonance point (2.93) the refractive index for this wave goes to infinity. Taking the thermal motion of the particles into account, as in the case of the upper hybrid resonance, clearly leads to conversion of this wave into a plasma wave. This conversion can be described approximately (as in Section 2.7) using a dispersion correction to ϵ_{xx}, which makes Eq. (2.104) take the form

$$(\epsilon - \beta'_{eff} N_x^2) N_x^2 + N_z^2 = 0, \tag{2.105}$$

where

$$\beta'_{eff} = \frac{3}{2} \frac{\omega_{pe}^2 + \omega_{Be}^2}{\omega_{Be}^2} \left(\frac{v_{Ti}^2}{c^2} + \frac{\omega^4}{4|\omega_{Be}|^3 \omega_{Bi}} \frac{v_{Te}^2}{c^2} \right)$$

and it has been noted that when $|\epsilon| \ll 1$ Eqs. (2.92) and (2.93) can be used. A comparison of Eqs. (2.83) and (2.105) shows that the conversion behavior in the neighborhoods of the lower and upper hybrid resonances is the same: when $\epsilon^2 \gg 4\beta_{eff}'N_z^2 |\eta|$ two modes with $N_x^2 \approx \epsilon/\beta_{eff}'$ and $N_x^2 \approx -N_z^2\eta/\epsilon$ exist in the plasma, and when $\epsilon^2 = 4\beta_{eff}'N_z^2 |\eta|$, the refractive indices of both modes are the same. The condition that the refractive indices of these modes coincide can be written as

$$\omega_{pc}^2 = \omega_{p\,LH}^2 \left[1 + \sqrt{6} N_z \frac{v_{Ti}}{c} \frac{\sqrt{|\omega_{Be}|^3 \omega_{Bi}}}{\omega^2 + |\omega_{Be}|\omega_{Bi}} \left(1 + \frac{\omega^4}{4|\omega_{Be}|^3 \omega_{Bi}} \frac{T_e}{T_i} \right)^{1/2} \right],$$

$$\tag{2.106}$$

where ω_{pc}^2 is the electron plasma frequency at the conversion point. Equation (2.106) determines the lower limit on the density in the plasma layer at which linear conversion of the slow wave into a plasma mode can occur.

Since the opacity region for the slow wave at the edge of a plasma with $\omega_{pe}^2 < \omega^2$ is very narrow, this wave is suitable for plasma heating, and most schemes for lower hybrid heating are based on using it. A slight disadvantage of this heating scheme is the need to excite waves that are slowed down along $\mathbf{B}_0$. Here, of course, it is desirable to use the minimum amount of slowing-down.

It has already been noted that Eqs. (2.102) and (2.104) are valid when $N_z^2 \gg 1$. It appears that when N_z is reduced the refractive indices of these waves converge at some point in the region with $\epsilon > 0$, and when $N_z < N_c$ (where N_c is some critical value), the slow mode undergoes linear conversion into the fast mode and vice versa (Fig. 2.14, curve 1). In the region between the conversion points, the discriminant of Eq. (2.25) (i.e., $A_2^2 - 4\epsilon A_0$) is negative, so that there the N_x^2 [roots of Eq. (2.25)] are complex quantities for both types of wave. An opacity region for these waves lies between the conversion points, and in order for plasma heating to be efficient, only waves with $N_z^2 > N_c^2$ can be used. The accessibility

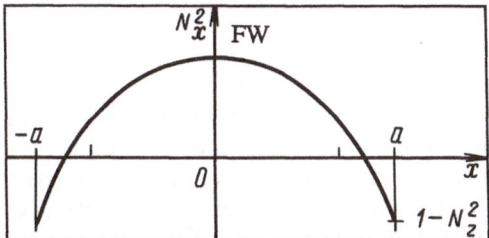

Fig. 2.15. The behavior of the square of the perpendicular refractive index of a fast wave (FW) in a plasma with a single ion species for $\omega \sim \omega_{Bi}$.

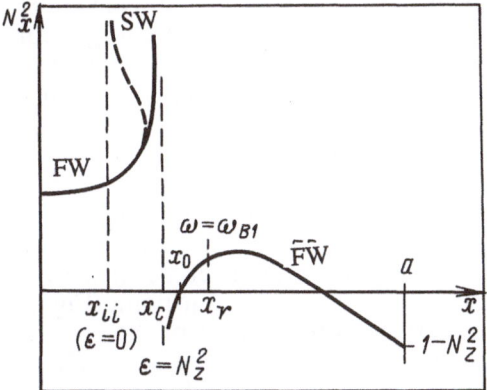

Fig. 2.16. The conversion of fast (FW) and slow (SW) waves in the neighborhood of a light ion cyclotron resonance: - - - slow wave; —— the solution of the approximate dispersion equation (2.102) for the fast wave.

of the lower hybrid resonance (i.e., the value of N_c) has been analyzed elsewhere [32, 42]. It turns out that the accessibility criterion can be written as

$$N_z^2 > 1 + (\omega_{pe}^2/\omega_{Be}^2)_{\text{LH}} \tag{2.107}$$

with enough accuracy for practical purposes [42].

2.9. PROPAGATION OF WAVES WITH FREQUENCIES $\omega \sim \omega_{Bi}$ AND $\omega \ll \omega_{Bi}$

We now consider how the propagation of waves in a plasma containing a single ion species changes as the frequency is lowered. According to Eq. (2.93), the lower hybrid resonance region is then displaced toward the edge of the plasma. At frequencies comparable to the ion cyclotron frequency, Eq. (2.90) is satisfied only in vary rarefied plasmas so that that the hybrid resonance is of no practical interest. The fast magnetosonic wave given by Eq. (2.102) is then most important. The cutoffs for this wave with $N_z > 1$ are determined by the condition

$$\omega_{pe}^2/(|\omega_{Be}|(\omega + \omega_{Bi})) = N_z^2 - 1. \tag{2.108}$$

The behavior of the refractive index $N_x(x)$ in a nonuniform plasma layer is illustrated schematically in Fig. 2.15. The maximum value of the transverse refractive index N_x is on the order of the Alfvén refractive index $N_A \sim \sqrt{4\pi n_i m_i c^2/B_0^2}$.

As noted in Section 2.8, an ion–ion hybrid resonance is possible in plasmas with different types of ions. In tokamaks this possibility is easily realized because of the nonuniformity of the magnetic field. We now discuss the effect of an ion–ion hybrid resonance on the propagation of the fast magnetosonic wave, using as an example a plasma with two ion species, hydrogen and deuterium. The principal changes associated with the existence of a second ion species in the plasma are easily understood if we assume that the ions of one type constitute a small minority. Thus, let hydrogen ions be present as a small additive and let the wave frequency be chosen so that the cyclotron resonance for the hydrogen ions ($\omega = \omega_{B1}(x_r)$) occurs inside the layer at the point $x = x_r$. Then, as can be seen from Eq. (2.95), the effect of the hydrogen ions will be important only in a small neighborhood of x_r when $\zeta \ll 1$. Outside of that neighborhood, the dispersion curve for the fast magnetosonic wave will be practically the same as that for a pure deuterium plasma (see Fig. 2.15).

At the cyclotron resonance point for the hydrogen ions ($x = x_r$), ϵ has a pole.[*] The conditions (2.95) for an ion–ion hybrid resonance are satisfied on the high-magnetic-field side near this point. Let us consider the behavior of the refractive index for the fast magnetosonic wave near this point (Fig. 2.16). Equations (2.91), (2.99), and (2.102) show that, as in the case of the electron cyclotron resonance, the refractive index has no singularities at the ion cyclotron resonance point in a cold plasma. Near this point, however, the inequality $\epsilon(x_c) = N_z^2$ is satisfied, so that Eq. (2.102) for N_x^2 goes to infinity. In addition, near $x = x_r$ there is yet another cutoff point for the wave. In fact, when two types of ion are present, the cutoffs are given by

$$\frac{\omega_{pe}^2}{|\omega_{Be}|(\omega + \omega_{B2})} \left[1 - \frac{\zeta(\omega_{B1} - \omega_{B2})}{\omega + \omega_{B1}} \right] = N_z^2 - 1 \tag{2.109}$$

and

$$\frac{\omega_{pe}^2}{|\omega_{Be}|(\omega - \omega_{B2})} \left[\frac{\zeta(\omega_{B1} - \omega_{B2})}{\omega_{B1} - \omega} - 1 \right] = N_z^2 - 1. \tag{2.110}$$

When $\zeta \ll 1$ the first of these differs little from Eq. (2.108) with $\omega_{Bi} = \omega_{B2}$, and the second is satisfied at some point $x = x_0$ which is near x_r when $\zeta \ll 1$. It can be shown that x_0 lies to the right of the point x_c corresponding to the singularity of Eq. (2.102), i.e., on the low magnetic field side.

---

[*]When the ion thermal motion is taken into account, ϵ_{xx} (as noted in Section 1.7) is finite. The following arguments, however, remain valid if the contribution of resonant ions to ϵ_{xx} is much greater than that of the background ions ($\omega_{p1}^2/\omega k_z v_{Ti} \gg \omega_{p2}^2/(\omega^2 - \omega_{B2}^2)$).

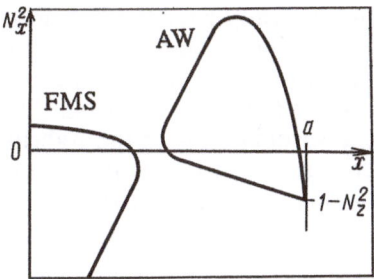

Fig. 2.17. The dispersion of waves at frequencies $\omega \ll \omega_{Bi}$. [AW and FMS denote the Alfvén and fast magnetosonic waves.]

It is important to note that according to Eq. (2.102), the pole of N_x^2 lies, not at the hybrid resonance point $\epsilon(x_{ii}) = 0$ as might be expected from the general analysis of Section 2.3, but at the point where $\epsilon(x_c) = N_z^2$. No contradiction arises here, however, if we recall that Eq. (2.102) is an approximation for the fast wave. Its going to infinity at $x = x_c$ means that the refractive index for the fast wave increases rapidly near there and approaches the refractive index for the slow mode (2.104), so that subdividing the waves into the modes (2.102) and (2.104) is no longer correct. The exact Eq. (2.25) implies that near $x = x_c$ these dispersion curves merge continuously into one another, a transition that is equivalent to linear conversion of the fast and slow waves. A clearer description of this transition can obtained by neglecting $(\epsilon - N_z^2)^2$ compared to g^2 in Eq. (2.25) and noting that $\epsilon \approx N_z^2 \ll |\eta|$, so that this equation takes the form

$$\epsilon N_x^4 - \eta (\epsilon - N_z^2) N_x^2 - \eta g^2 = 0. \tag{2.111}$$

This equation is of the same type as Eq. (2.83). The conversion point for the waves corresponds to the condition

$$N_z^2 - \epsilon = 2N_z (\omega/\omega_{pe}) |g|. \tag{2.112}$$

When $|N_z^2 - \epsilon| \gg 2N_z(\omega/\omega_{pe}) |g|$, Eq. (2.111) has two approximate solutions for N_x^2:

$$N_x^2 \simeq (\eta/\epsilon)(\epsilon - N_z^2) \quad \text{and} \quad N_x^2 \simeq -g^2/(\epsilon - N_z^2),$$

which correspond to the slow and fast waves. At the conversion point the refractive indices of these waves coincide and are given by

$$N_x^2 = \omega_{pe} |g|/(\omega N_z). \tag{2.113}$$

The refractive index has a singularity only at the hybrid resonance point for the slow wave. The variation in N_x^2 when linear conversion of the fast and slow waves is taken into account is shown in Fig. 2.16 by the dashed curve.

We note that under actual conditions the refractive index of the slow wave is often so large that spatial dispersion must be taken into account in the fast and slow wave conversion region, and not only near the hybrid resonance $\epsilon = 0$. Then the detailed variation in the refractive index for the slow wave may also change.

Finally, we consider the propagation of waves with $N_z^2 \gg 1$ when $\omega < \omega_{Bi}$. As can be seen from Eqs. (2.90) and (2.91), in this case ϵ is positive for arbitrary densities, so that the phenomena associated with the hybrid resonances do not occur. As in the previous case, here the waves described by Eqs. (2.102) and (2.104) are also of interest. When $\omega \ll \omega_{Bi}$, the equation for the slow mode (2.104) can be transformed to

$$\omega^2 = k_z^2 \, v_A^2 \, (1 + k_x^2 v_A^2 / \omega_{Bi} |\omega_{Be}|)^{-1}, \tag{2.114}$$

where $v_A = c\omega_{Bi}/\omega_{pi}$ is the Alfvén speed. The wave described by the dispersion equation (2.114) with $k_x^2 v_A^2/(|\omega_{Be}|\omega_{Bi}) \ll 1$ is known as the Alfvén wave.

When $\omega/\omega_{Bi} \ll 1$, Eq. (2.102) is also considerably simplified, since then $g \simeq (\omega/\omega_{Bi})\omega_{pi}^2/\omega_{Bi}^2 \ll \epsilon \simeq \omega_{pi}^2/\omega_{Bi}^2$ and everywhere, except in a small neighborhood of the point x_c ($\epsilon(x_c) = N_x^2$), it is possible to write $N_x^2 = \epsilon - N_z^2$. This yields the standard equation for the fast magnetosonic wave

$$\omega^2 = k^2 \, v_A^2. \tag{2.115}$$

The neighborhood of the point where $\epsilon(x_c) = N_z^2$ (i.e., $\omega^2 = k_z^2 v_A^2$) requires closer attention. Equation (2.102) for N_x^2 has a pole at this point and goes to zero on its left and right at the points $(\epsilon - N_z^2)^2 \simeq (\omega^2/\omega_{Bi}^2)\epsilon^2$. The transverse refractive index of the slow wave (2.104) goes to zero at the singularity of the fast wave, $x = x_c$. Since Eqs. (2.102) and (2.104) are approximations, it is clear, as before, that linear conversion of the fast magnetosonic wave into the Alfvén wave occurs in the neighborhood of this point. The dispersion curves for this case are illustrated schematically in Fig. 2.17. As before, the thermal motion of the particles has a strong effect on the conversion behavior.

2.10. EFFECT OF REAL PLASMA INHOMOGENEITIES ON THE PROPAGATION OF WAVES IN TOROIDAL SYSTEMS

In Sections 2.4–2.9 we have discussed the propagation and absorption of waves in toroidal systems using a one-dimensional model of an inhomogeneous plane plasma layer in the approximation of geometric optics. Here we examine the limits of applicability of this approach in somewhat more detail and consider the complications which arise in attempting to take more realistic account of the system geometry and of the inhomogeneities in the plasma and magnetic field. First, we briefly discuss why the approximation of geometric optics may be unsuitable in the simple form used earlier.

It is well known that the approximation of geometric optics fails near turning points ($N_x^2 = 0$) and wave conversion points ($N_{x1}^2 = N_{x2}^2$). The wave field near such points cannot be represented in the quasiclassical form (2.37) and the wave equations must be analyzed directly. The situations where geometric optics fails in this way have been studied thoroughly and classified [30, 37, 38]. Standard equations have been derived for the structure of the field in the neighborhood of turning points and in wave conversion regions. The physics of the wave processes that occur here is completely clear and has been described previously.

Certain difficulties arise in describing the propagation of waves through a cyclotron resonance region in a hot plasma because the dielectric tensor is highly non-Hermitian when $\omega \to |\omega_B|$. Furthermore, the very concept of the dielectric tensor as a function of the frequency ω and wave vector k is strictly applicable only for a uniform magnetic field, and in order to evaluate the applicability of geometric optics it is necessary to examine the complete system (2.1), (2.4), and (2.5), including both the field equations and the kinetic equation. In other words, these difficulties are related to the nonlocal character of the coupling between the current in the plasma and the wave field which induces this current. Various aspects of this problem have been discussed in reviews [43–46] which include references to original work on this topic. Without going into detail, we mention several physical phenomena which arise during the passage of a wave through a cyclotron resonance region. First, there is an asymmetry in the propagation of waves along and opposite to the gradient in the magnetic field: when a wave is incident on the resonance layer from the low field side, part of its energy is reflected, while when a wave is incident from the high field side it can pass through the resonance layer without reflections. There are also some extremely interesting effects in which wave barriers become transparent [38, 47]. During plasma heating in toroidal systems, these effects appear to be unimportant, but they may lead to some quantitative changes.

In experiments on small devices at low frequencies the wavelength in the plasma may be comparable to the plasma radius. Under these conditions the plane model is, of course, completely inapplicable. In the absence of cyclotron resonance effects, a cylindrical model for the plasma column, in which the plasma parameters depend only on the radius and the magnetic field is uniform, is closer to reality. Then the wave field must be represented by

$$\mathbf{E}(\mathbf{r}, t) = \mathbf{E}(r) \exp(im\vartheta + ik_z z - i\omega t) \tag{2.116}$$

rather than by Eq. (2.37). The closure of the system in the toroidal direction can be taken into account qualitatively by setting $k_z = l/R_0$, where l is an integer equal to the number of waves that fit along the major circumference of the torus.

Substituting Eq. (2.116) into the system (2.1) and using Eq. (2.21) for the dielectric tensor, in some cases it is possible to obtain comparatively simple equations for the behavior of the field in the plasma. Thus, setting $m = 0$ and considering fast magnetosonic waves (for which we may assume $E_z \approx 0$, as in the plane case), we find

$$\frac{d^2 E_\vartheta}{dr^2} + \frac{1}{r} \frac{dE_\vartheta}{dr} + (k_r^2 - \frac{1}{r^2}) E_\vartheta = 0, \tag{2.117}$$

where

$$k_r^2 = \frac{\omega^2}{c^2} \left[\epsilon - N_z^2 - \frac{g^2}{\epsilon - N_z^2} \right] = \frac{\omega^2}{c^2} N_r^2, \quad N_z = \frac{lc}{\omega R_0}, \quad E_r = \frac{-ig E_\vartheta}{\epsilon - N_z^2}.$$

When $r = 0$, the solutions of this equation satisfy the condition of finiteness. At the plasma boundary $r = a$ (the limiter radius) the functions E_r and E_ϑ are matched to the corresponding vacuum solutions of Eqs. (2.1) with the aid of the usual boundary conditions $\epsilon E_r + igE_\vartheta|_{r \to a-0} = E_r|_{r \to a+0}$ and $E_\vartheta|_{r \to a-0} = E_\vartheta|_{r \to a+0}$. The boundary conditions at an ideally conducting vessel wall ($r = a_c > a$) are that the tangential components of the electric field and the normal components of the magnetic field of the wave go to zero.

Solving Eq. (2.117) with these boundary conditions yields the eigenfrequencies and eigenfunctions of resonators formed by plasma cylinders of length $2\pi R_0$ with overlapping ends [i.e., $\mathbf{E}(z) = \mathbf{E}(z + 2\pi R_0)$]. Presumably, the results for weak toroidicity ($a/R_0 \ll 1$) would not differ greatly from the eigenfrequencies of a real toroidal resonator. In order to avoid complicating the discussion with details that are not fundamental, we shall neglect the vessel walls and consider the simpler problem of finding the eigenfrequencies of a plasma cylinder whose density is not zero at $r = a$, but has a small value $n_e(a)/n_e(0) \ll 1$ and falls off monotonically as $r \to \infty$. We shall also assume that $\epsilon - N_z^2$ is not zero anywhere in the plasma. The eigenfrequencies can then be found using the simple rules for quasiclassical quantization. Introducing the new independent variable $\xi = \ln (\omega r/c)$ in Eq. (2.117), we reduce it to the standard form

$$d^2 E_\vartheta/d\xi^2 + [N_r^2 \exp(2\xi) - 1] E_\vartheta = 0.$$

According to the general rules [48], the eigenfrequencies ω_{ls} will be given by

$$\int_{\xi_1}^{\xi_2} \sqrt{N_r^2 \exp(2\xi) - 1} \, d\xi = \frac{\omega}{c} \int_{r_1}^{r_2} \sqrt{N_r^2 - \frac{c^2}{\omega^2 r^2}} \, dr = \pi (s + \frac{1}{2}), \tag{2.118}$$

where the points r_1 and r_2 ($r_1 < r_2$) are found from the condition that the expression under the radical goes to zero. When $s \gg 1$ (i.e., $\overline{N}_r^2(\omega^2 a^2/c^2) \gg 1$), the term $c^2/(\omega^2 r^2)$ under the integral (which is a consequence of the cylindrical geometry) can be neglected and Eq. (2.118) takes the very simple form

$$(\omega/c) \int_0^{r_0} N_r dr = \pi (s + 1/2), \tag{2.119}$$

where $N_r(r_0) = 0$.

The eigenfrequencies of the plasma column in the low-frequency region $\omega \ll |\omega_{Be}|$ are usually called the magnetosonic resonance frequencies [49]. Equations (2.118) and (2.119) are approximations for finding these frequencies when the effect of the vessel walls is unimportant. Clearly, the presence of a conducting wall near the plasma boundary does not qualitatively change the resonance behavior, but causes a shift in the eigenfrequencies which depends on the depth of the opacity region at the plasma edge (with $r > r_2$). When $(\omega/c) \int_{r_2}^{a_c} \sqrt{c^2 \omega^{-2} r^{-2} - N_r^2} \, dr \gg 1$ this shift is exponentially small.

These results are applicable only when conversion of the fast magnetosonic wave into the slow mode does not occur in the plasma, i.e., $\epsilon - N_z^2$ does not go to zero anywhere in the plasma. The complications which arise in the presence of wave conversion depend strongly on the specific conditions. If it is possible to neglect the toroidicity of the magnetic field even as an approximation and assume that $\epsilon - N_z^2$ goes to zero for a fixed value of r in the minor cross section of the plasma, then the effect of conversion with $\omega \leq \omega_{Bi}$ can be taken into account by the above-mentioned formal approach of introducing an effective dissipation [$\epsilon \rightarrow \epsilon + i\nu_{eff}/\omega$)] which can be neglected in the final formulas. Because of the presence of a singularity in the coefficients of Eq. (2.117), when $\epsilon - N_z^2 = 0$ the effective absorption of fast magnetosonic waves, which models their conversion into the slow mode, remains finite even in the limit of vanishingly small dissipation ν_{eff}/ω. The method of solving Eq. (2.117) then no longer reduces to the simple quasiclassical quantization rule (2.118); however, the main consequence of wave conversion can be understood without any computations. It involves a reduction in the Q-factor of the magnetosonic cavities. When the conversion efficiency is high and a significant fraction of the energy in the fast wave goes into the slow mode during a single "pass" over the minor radius, this reduction in the Q-factor is so great that it is no longer meaningful to speak of a resonance. A qualitative idea of the efficiency of conversion under different conditions can be obtained on the basis of our earlier discussion, which allows us to identify situations in which magnetosonic resonances with a high Q-factor exist.

The changes in the wave behavior are more fundamental when the toroidal inhomogeneity of the magnetic field is significant. This happens, for example, in plasmas with a small admixture of light ions for which for a cyclotron resonance may exist inside the plasma column. Then the equality $\epsilon = N_z^2$ is satisfied (in the first approximation) for a fixed distance from the symmetry axis of the torus regardless of the minor radius. The plasma inhomogeneity is fundamentally two-dimensional, so that, even as an approximation, it is impossible to treat the wave field as a function of the poloidal angle ϑ of the form $\exp(im\vartheta)$ with $m = \text{const}$. The situation is still more complicated when the poloidal magnetic field is taken into account and the system becomes inhomogeneous along the lines of force of the total magnetic field. Then $N_\parallel$, the refractive index parallel to the magnetic field, cannot be regarded as a conserved quantity. This makes the problem much more complicated even in the approximation of geometric optics.

A natural generalization of Eq. (2.37) to the case of a multidimensional inhomogeneous medium is given by

$$E(r) = E_0(r) \exp[i\psi(r) - i\omega t], \qquad (2.120)$$

where the vectors $E_0(r)$ and $\nabla\psi(r)$, as with a one-dimensional inhomogeneity, are regarded as slowly varying functions of the coordinates compared to $\exp[i\psi(r)]$. Hence, when Eq. (2.120) is substituted into the wave equations, only the exponent has to be differentiated in the zeroth approximation. This again yields a dispersion equation of the form (2.25), but now with two unknown functions $N_\perp^2(r) = c^2[\nabla\psi B_0]^2/B_0^2\omega^2$ and $N_\parallel^2(r) = c^2(\nabla\psi B_0)^2/B_0^2\omega^2$. Here difficulties arise even in defining the resonance region ($N^2 \rightarrow \infty$) for the waves. Indeed, in a one dimensionally inhomogeneous medium the refractive index goes to infinity only because of its component in the direction of the inhomogeneity. This immediately determines the angle between the wave vector k and the magnetic field B_0 on which the location of the resonance depends [Eq. (2.33)]. For a two-dimensional inhomogeneity the resonance behavior is much more complicated.

When the wavelength in the plasma is much shorter than the dimensions of the system, many of the problems originating in the complex character of the plasma inhomogeneity can be solved in the approximation of geometric optics (2.120) by analyzing the ray trajectories of the waves [30, 32, 43]. We now illustrate the basic ideas of this approach, using as an example a two-dimensional problem in which there are only two projections of the "wave vector" $k_x = \partial\psi/\partial x$ and $k_z = \partial\psi/\partial z$. We shall also assume that the plasma parameters depend only on x and z. We write the dispersion equation relating k_x, k_z, and the wave frequency ω in the symbolic form

$$D\ (k_x,\ k_z,\ \omega) = 0. \tag{2.121}$$

In the special case of a cold plasma in a uniform magnetic field $\mathbf{B_0}$ along the z axis, Eq. (2.121) is the same as Eq. (2.25), so that $D(k_x, k_z, \omega)$ is a fourth-degree polynomial in k_x and k_z. When the magnetic field is inhomogeneous with curved lines of force, Eq. (2.121) may be reduced to the form (2.25) by local rotation of the coordinate system.

The zeroth-order approximation of geometric optics for a two-dimensional inhomogeneity of the medium, therefore, reduces to determining two functions $k_x(x, z)$ and $k_y(x, z)$ which are related by Eq. (2.121). This can be done if we note that, because $\mathbf{k} = \nabla\psi$, the following relation holds:

$$\partial k_x/\partial z = \partial k_z\ /\ \partial x. \tag{2.122}$$

We then differentiate Eq. (2.121) with respect to x and z, noting that $D(k_x, k_z, \omega)$ depends directly on the coordinates because of the dependence of k_x and k_z on position:

$$\frac{\partial D}{\partial x} + \frac{\partial D}{\partial k_x}\frac{\partial k_x}{\partial x} + \frac{\partial D}{\partial k_z}\frac{\partial k_z}{\partial x} = 0;$$

$$\frac{\partial D}{\partial z} + \frac{\partial D}{\partial k_x}\frac{\partial k_x}{\partial z} + \frac{\partial D}{\partial k_z}\frac{\partial k_z}{\partial z} = 0.$$

Using Eq. (2.122), we can transform this equation to

$$\frac{\partial D}{\partial \mathbf{r}} + \left(\frac{\partial D}{\partial \mathbf{k}}\nabla\right)\mathbf{k} = 0. \tag{2.123}$$

It is easy to see that Eq. (2.123) does not change its form on going to the three-dimensional problem. Its three projections onto the coordinate axes yield three partial differential equations in the functions k_x, k_y, and k_z. Solving these equations gives the desired answer.

It is often convenient to examine the system of ordinary differential equations that is equivalent to Eq. (2.123) rather than this equation itself. That system can be obtained as follows. We differentiate Eq. (2.121) with respect to the projections of the vector $\mathbf{k}$, noting that the frequency ω of the wave which satisfies Eq. (2.121) is a function of $\mathbf{k}$:

$$\frac{\partial D}{\partial \mathbf{k}} + \frac{\partial D}{\partial \omega} \mathbf{v}_g = 0, \tag{2.124}$$

where $\mathbf{v}_g = d\omega/d\mathbf{k}$ is the group velocity vector of the wave. The equation

$$\frac{d\mathbf{r}}{dt} = \mathbf{v}_g = -\frac{\partial D}{\partial \mathbf{k}} \left[\frac{\partial D}{\partial \omega} \right]^{-1} \tag{2.125}$$

determines the ray trajectory of the wave along which its energy propagates at velocity v_g. Using Eqs. (2.125) and (2.124), we can write Eq. (2.123) in the form

$$\frac{d\mathbf{k}}{dt} = \frac{\partial D}{\partial \mathbf{r}} \left[\frac{\partial D}{\partial \omega} \right]^{-1}, \tag{2.126}$$

where the quantity $\mathbf{r}$ on the right-hand side denotes the coordinate $\mathbf{r}(t)$ of the ray trajectory corresponding to the solution of Eq. (2.125).

The system of Eqs. (2.125) and (2.126) can thus be used to find the ray trajectories of the waves and to calculate the change in the wave vector along these trajectories.

2.11. DISTORTION OF THE DISTRIBUTION FUNCTIONS OF CHARGED PARTICLES IN THE FIELD OF A MONOCHROMATIC WAVE

Until now we have examined the propagation and absorption of waves in the linear approximation. The action of a strong electromagnetic wave on a plasma, however, can lead to nonlinear effects. The first of these effects is distortions in the distribution function of the particles, for which the mechanism is already obvious from an examination of collisionless energy absorption in plasmas (see Section 2.6). Indeed, a wave causes large changes in the motion of resonant particles, whose velocity obeys the condition

$$\omega - s\omega_B - k_z v_z = 0, \quad s = 0, \pm 1, \pm 2 \ldots \tag{2.127}$$

The first-order correction to the unperturbed distribution function caused by the wave has a pole at a velocity determined by this condition.

It is natural to expect noticeable deviations of the distribution function from Maxwellian at velocities close to the resonant velocities when a more accurate nonlinear treatment is used. The physics of these deviations becomes evident when it is noted that under real conditions a plasma is heated in the following way: the wave energy is absorbed as a result of one or another collisionless mechanism by the resonant particles. Collisions tend to Maxwellianize the distribution function, "collecting" energy from the resonant particles and passing it on to all the particles in the plasma. Clearly, when the range of resonant velocities is small, the distribution function must differ substantially from Maxwellian in order for a significant flux of energy from the resonant particles to the background particles (i.e., for rapid heating) to develop. The distortions in the distribution function should be much smaller in the opposite case, when resonant absorption is "smeared out" over most of the distribution function. Heating of this sort occurs, for example, when waves are excited over a broad spectrum in $N_{\parallel}$ or in an inhomogeneous magnetic field.

Large distortions in the distribution function are undesirable since they can lead to plasma instabilities. Sometimes, however, they play a positive role, as when they are deliberately produced during rf current drive.

The evolution of the particle distribution functions under the influence of waves excited in a plasma is the subject of an extensive literature. The basic results on this topic can be found in various monographs and reviews [29, 32, 44, 51, 52–55]. Rather than discuss these results in detail, we shall only examine qualitatively the basic physical phenomena of importance for understanding the processes that occur during rf heating of plasmas. In order to clearly illustrate the evolution of the distribution function during the application of rf waves, we begin the discussion with the simplest case, nonlinear absorption of plasma (Langmuir) waves propagating in a uniform plasma along a constant external magnetic field. We begin with the kinetic equation (2.3), substituting into it an electron distribution function of the form $f = F + f_1 + f_2 + \ldots$, where $F(\mathbf{v}, t)$ is the smooth part of the distribution function formed by the wave, f_1 describes the oscillations at the wave frequency in the lowest order in the wave field, and f_2 describes the oscillations at the harmonic frequency 2ω. When the wave amplitude is sufficiently low, the distribution function $F(\mathbf{v}, t)$ evolves rather slowly, so that in the first approximation the wave field can be written in the form

$$E\,(z,\,t) = E_0\,(t)\,\exp\left[i\,(kz - \omega t)\right],\tag{2.128}$$

where $E_0(t)$ is a smoothly varying function. In higher approximations field components that oscillate at the combination frequencies $s\omega$ appear, but we shall assume that they are small.

The function f_1 obeys the equation

$$\frac{\partial f_1}{\partial t} + v_z\,\frac{\partial f_1}{\partial z} + \frac{e}{m_e}\,E(z,\,t)\,\frac{\partial F}{\partial v_z} = \frac{\delta f_1}{\delta t}.\tag{2.129}$$

This is the same as Eq. (2.5) for the linear theory with the initial Maxwellian distribution function $f_0(\mathbf{v})$ replaced by $F(\mathbf{v})$. In this approximation the wave field $E(z, t)$ satisfies the linear equations (2.26) and (2.80) in which the function $F(\mathbf{v})$ was used for calculating ϵ_{zz}. Since significant deviations in $F(\mathbf{v})$ can be expected only near velocities v_z close to the phase velocity $v_{\mathrm{ph}} = \omega/k$ of the wave, the real part of ϵ_{zz} (which is determined by the bulk of the distribution for $\omega^2 \gg k^2 v_{Te}^2$) remains practically unchanged, so that the real part of the frequency ω is given, as before, by Eq. (2.81). In accordance with Eq. (2.57), the damping decrement of the wave has the form

$$\gamma = \frac{\pi}{2}\,\frac{\omega_p^3}{n_e k^2}\,\frac{\partial F_z}{\partial v_z},\tag{2.130}$$

where $F_z = 2\pi \int\limits_0^{\infty} F(\mathbf{v})v_{\perp}dv_{\perp}$. Then the amplitude $E_0(t)$ obeys

$$dE_0/dt = \gamma E_0.\tag{2.131}$$

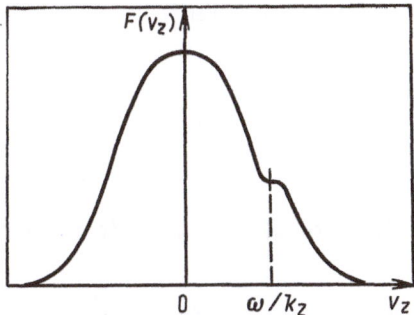

Fig. 2.18. The distortion of the electron distribution function in the field of a monochromatic wave.

In order to obtain an equation for the evolution of $F(\mathbf{v})$ in the field of a wave, the nonoscillatory part of the nonlinear term $E \partial f / \partial \mathbf{v}$ must be kept in Eq. (2.129):

$$\frac{\partial F}{\partial t} + \mathbf{v}\,\frac{\partial F}{\partial \mathbf{r}} + \frac{e}{m_e}\left\langle \operatorname{Re} E \operatorname{Re} \frac{\partial f_1}{\partial \mathbf{v}} \right\rangle = \frac{\delta F}{\delta t}, \qquad (2.132)$$

where the sign $<\ >$ denotes averaging over the fast oscillations. To obtain the nonlinear term in Eq. (2.132) in explicit form, a specific collisional term must be chosen in Eq. (2.129). Particle collisions in a fully ionized plasma are determined by the extremely complicated Landau collision integral [29, 31]. Since our purpose is not to obtain quantitative results, but to clarify the qualitative behavior, we shall use a much simpler and widely used collision integral of the form (relaxation model)

$$\delta f / \delta t = -\nu_{\text{eff}}\,(f - f_0), \qquad (2.133)$$

which describes the relaxation of the perturbed distribution function over a characteristic time $1/\nu_{\text{eff}}$ on the order of the collision time. Substituting Eq. (2.128) in Eq. (2.129) and taking note of Eq. (2.133), we find

$$f_1 = -\frac{ie}{m_e}\,\frac{\partial F}{\partial v_z}\,\frac{E}{\omega - k v_z + i\nu_{\text{eff}}}.$$

Using this expression for calculating the nonlinear term in Eq. (2.132), we obtain

$$\frac{\partial F}{\partial t} + v_z\,\frac{\partial F}{\partial z} - \frac{\partial}{\partial v_z}\,D_{\text{hf}}\,\frac{\partial F}{\partial v_z} = -\nu_{\text{eff}}\,(F - f_0), \qquad (2.134)$$

where

$$D_{\text{hf}} = \frac{e^2\,EE^*}{2m_e^2}\,\frac{\nu_{\text{ef}}}{(\omega - k v_z)^2 + \nu_{\text{eff}}^2}. \qquad (2.135)$$

We shall assume that the amplitude of the plasma wave is kept constant by an external source. Then the steady-state solution of Eq. (2.134) can be examined.

When the field amplitude E_0 is small enough, Eq. (2.134) can be solved by perturbation theory, setting $F = f_0 + \delta f$, where $|\delta f| \ll f_0$. Then, neglecting the derivatives of slowly varying functions of the velocity compared to those of the rapidly varying ones where $(\omega - kv_z)^2 \lesssim v_{eff}^2$ in the resonance denominator when $v_{eff} \ll \omega$, we obtain

$$\delta f = \frac{e^2 E E^*}{m_e^2 v_{Te}^2} \frac{k^2 v_z^2 - \omega^2 - v_{eff}^2}{[(\omega - kv_z)^2 + v_{eff}^2]^2} f_0. \tag{2.136}$$

The derivative of this function at $v_z = v_{ph} \equiv \omega/k$ is positive and equals

$$\frac{\partial}{\partial v_z} \delta f \bigg|_{v_z = v_{ph}} = \frac{2\omega^2 e^2 |E|^2}{v_{Te}^2 v_{eff}^4 v_{ph} m_e^2} f_0, \tag{2.137}$$

so that the nonlinear action of the wave leads to a reduction in the damping decrement γ that is proportional to $\partial F/\partial v_z$. This reduction is unimportant when the following criterion is satisfied:

$$-\frac{\partial \delta f}{\partial v_z} \left[\frac{\partial f_0}{\partial v_z} \right]^{-1} = \frac{e^2 k^2 |E|^2}{m_e^2 v_{eff}^4} \ll 1. \tag{2.138}$$

Although condition (2.138) also limits the applicability of the perturbation theory approach described here for finding F, it is clear from qualitative considerations that when this condition is not satisfied, the perturbations in the distribution function in the narrow region $(\omega - kv)^2 \lesssim v_{eff}^2$ are important. In the formal limit $e^2 k^2 E E^*/(m_e^2 v_{eff}^4) \to \infty$ a plateau develops in the distribution function and the decrement goes to zero (Fig. 2.18). The energy density p_E transferred from the wave to the plasma per unit time can also be written in the form

$$p_E = -2\pi m_e \int v_z D_{hf} \frac{\partial F}{\partial v_z} v_\perp dv_z dv_\perp. \tag{2.139}$$

To evaluate this integral we note that the function $v_{eff}[(\omega - kv_z)^2 + v_{eff}^2]\pi^{-1}$ in D_{hf} approaches the δ-function $\delta(\omega - kv_z)$ as $v_{eff} \to 0$. For the integration over v_z it can be replaced by the δ-function for finite but small values of v_{eff} as well. Then we obtain the reasonable solution

$$p_E = 2\gamma E E^*/(8\pi), \tag{2.140}$$

where γ is the damping decrement for the wave given by Eq. (2.130) and $EE^*/(8\pi)$ can be interpreted as the energy density in the plasma wave.

These results make it possible to view Landau damping from yet another standpoint. It is clear from our discussion that in the steady state the linear approximation is valid only when the criterion (2.138) is satisfied. Although it is evident that the damping decrement is then practically independent of the collision frequency, the absorption of the wave energy is completely determined by collisions. This independence of the absorbed energy from the collision frequency is related to the fact that collisions determine the energy stored by the resonant electrons in the wave field. When the collision frequency is lowered, the stored energy increases to the extent necessary for maintaining constant collisional energy transfer from the resonant electrons to the bulk plasma.

The applicability of these results to the steady state is limited by the criterion (2.138). It is clear, however, that the formulas of the linear theory are valid for the evolution in a given initial field, regardless of condition (2.138), until times when nonlinear distortions appear in the distribution function. In order to obtain an idea of this time scale, we now examine in more detail the motion of electrons in the field (2.128) of a monochromatic wave. For this purpose it is convenient to transform to a coordinate system moving at the phase velocity v_{ph} of the wave. In this system the problem reduces to determining the motion of an electron in an electrostatic field with a potential $\varphi(z) = \Phi_0 \sin kz$. The electron's motion in such a field is determined by the total energy $\mathscr{E} = mv_z^2/2 - |e| \varphi(z)$. When $\mathscr{E} < |e| \Phi_0$, the electrons are trapped in a potential well, and when $\mathscr{E} > |e| \Phi_0$, they move without reflection from potential barriers. To estimate the characteristic oscillation period of electrons trapped in the potential well, we consider their motion near one of the potential maxima $\mathscr{E} \to -|e| \Phi_0$ with the aid of the equation

$$m_e \ddot{z}_1 = |e| k \Phi_0 \cos k (z_0 + z_1) \simeq -|e| k^2 z_1 \Phi_0,$$

where z_1 is the displacement from the position of the potential maximum, $z_0 = \pi/(2k)$. Here z_1 obeys the harmonic oscillator equation with a frequency $\omega_b = \sqrt{k |eE|/m_e}$. In general the period τ_b of these oscillations differs little from ω_b^{-1} for most of the particles.

The motion of the resonant electrons, whose velocity in the laboratory coordinate system obeys

$$(v_z - v_{ph})^2 < 2 |e| \Phi_0/m_e, \tag{2.141}$$

is substantially nonlinear. It is natural to choose a time τ_b equal to ω_b^{-1} as the characteristic time for this nonlinear motion to develop. This time, of course, also determines the characteristic time for the appearance of nonlinear distortions in the distribution function. Since the period of the trapped electrons depends on their energy $\mathscr{E}$, their phases shift over several periods and the distribution function develops a plateau over the whole range of velocities determined by Eq. (2.141). Then at any given time the numbers of particles accelerated by the field and slowed down in it become equal and absorption of the wave energy ceases.

Collisions can greatly change this picture only when

$$\tau_b \nu_{eff} \gg 1. \tag{2.142}$$

Since an electron is accelerated anew by the wave field after every collision, nonlinear oscillations cannot develop under conditions (2.142) and the equations of the linear theory will be applicable. In particular, the region with collisionally induced nonlinear distortions in the distribution function $(v_z - v_{ph})^2 \lesssim \nu_{eff}^2 v_{ph}^2/\omega^2$ [Eq. (2.136)] will be considerably larger than the region (2.141) for trapping of particles in the potential well.

It is interesting to note that the criteria (2.138) and (2.142) are the same. This means that for a monoenergetic wave the above quasilinear approach, in which the oscillating part of the distribution function obeys the "linear" equation (2.129) and only the averaged effect of the wave on the smooth part of the distribution function F is taken into account (to within terms of order $|E|^2$), is applicable only as long as F deviates negligibly from Maxwellian. At first glance this conclusion seems surprising, since the condition $|f_1| \ll F$ is satisfied when the field E obeys a much weaker condition than Eq. (2.138), i.e.,

$$|\dot{e}E|\nu_{ph}/(m_e\nu_{eff}v_{Te}^2) \ll 1.$$

However, there is no contradiction here. Using Eq. (2.3) to calculate the second-order correction to the distribution function f_2, we find that it contains terms of the form

$$f_2 \approx \frac{e^2|E|^2k}{m_e[(\omega-kv_z)+i\nu_{eff}]^2[2(\omega-kv_z)+i\nu_{eff}]}$$

Evidently $|f_2| \ll |f_1|$ only when Eq. (2.138) is satisfied. Similarly, the power of the resonance denominator increases by two for each subsequent correction in the field, so that Eq. (2.138) is absolutely necessary for convergence of the perturbation theory expansion of the distribution function.

This model collision integral does not reflect one of the principal properties of Coulomb collisions: the diffusive variation in the velocity vector of the particles owing to repeated small angle scattering. Strictly speaking, therefore, our treatment applies to a weakly ionized plasma. Taking Coulomb collisions into account leads to some quantitative changes in the picture described above. In particular, Eq. (2.142) for the limit of applicability of the linear theory changes significantly. In fact, the phase of the nonlinear oscillations of the trapped particles can be shifted by changing their velocity $v_z \approx v_{ph}$ by a small amount, on the order of $\sqrt{2e\Phi_0/m_e}$. Because of the diffusive character of Coulomb collisions, the time for this change is considerably smaller than the characteristic time for 90° scattering, $1/v_c$, and is on the order of $|e|\Phi_0/(v_cm_ev_{ph}^2)$. Given that this time must be less than the oscillation period τ_b, we obtain the inequality

$$|eE|^3/(km_e^3v_{ph}^4v_c^2) \ll 1. \qquad (2.143)$$

We now examine the absorption of waves through Coulomb collisions in more detail using a model for the collision integral of the form

$$\frac{\delta f}{\delta t} = \frac{\partial}{\partial v_z}D_c\left(\frac{m_ev_z}{T_e}f+\frac{\partial f}{\partial v_z}\right), \qquad (2.144)$$

where $D_c = v_{ph}^2v_c$, and v_c is the frequency of Coulomb collisions for a particle with a velocity $v_z = v_{ph} \gg v_{Te}$. Substituting this expression in Eq. (2.129) and transforming to the new variable $\hat{u} = (v_z - v_{ph})(\omega/v_cv_{ph}^3)^{1/3}$, for the steady state we obtain

$$d^2f_{10}/d\hat{u}^2 - i\hat{u}f_{10} = H + O[(v_c/\omega)^{1/3}]$$

in the lowest order in $(v_c/\omega)^{1/3}$ ($\partial F/\partial t = \partial F/\partial z = 0$), where f_1 is represented by

$$f_1 = f_{10} \exp [i(kz - \omega t)], \quad H = \frac{e E_0}{m_e (v_c \omega^2)^{1/3}} \frac{\partial F}{\partial v_z} \bigg|_{v_z = v_{ph}}.$$

The solution of this equation that falls off as $|\hat{u}| \to \infty$ has the form

$$f_{10} = H \int_0^\infty \exp [i(\hat{u}t + it^3/3)] dt. \tag{2.145}$$

It is easy to see that when $\hat{u} = 0$ the integral in Eq. (2.145) is on the order of unity, while when $\hat{u}^2 \gg 1$ (but with $(v_z - v_{ph})^2 \ll v_{Te}^2$) the main contribution to the integral is from the beginning of the integration path, so that

$$f_{10} \simeq \frac{iH}{\hat{u}} = -\frac{ie E_0}{m_e (\omega - kv_z)} \frac{\partial F}{\partial v_z}, \tag{2.146}$$

where it is assumed that the characteristic scale for changes in F is v_{Te}, so that the difference between $\partial F/\partial v_z$ and $(\partial F/\partial v_z)|_{v=v_{ph}}$ can be neglected in this range of velocities. In the region with $(v_z - v_{ph})^2 \geq v_{Te}^2$ collisions can be taken into account by perturbation theory. As a result, in the zeroth approximation we again obtain Eq. (2.146) but now for arbitrary velocities v_z as well. Using Eq. (2.145), we now find an equation for F that is an analog of the steady-state equation (2.134):

$$\frac{\partial}{\partial v_z} D_{hf} \frac{\partial F}{\partial v_z} + \frac{\partial}{\partial v_z} D_c \left(\frac{m_e v_z}{T_e} F + \frac{\partial F}{\partial v_z} \right) = 0, \tag{2.147}$$

where

$$D_{hf} = \frac{e^2 |E|^2}{m_e^2 (v_c \omega^2)^{1/3}} \int_0^\infty \cos \hat{u}t \exp (-t^3/3) dt.$$

After integrating once with respect to the velocity we obtain

$$\frac{\partial F}{\partial v_z} + \frac{D_c}{D_{hf} + D_c} \frac{m_e v_z}{T_e} F = 0.$$

The solution of this equation can easily be written in quadrature form. Here, where the deviations of F from Maxwellian are small compared to the deviations in their derivatives, however, $m_e v_z/T_e$ can be replaced by the derivative of the Maxwellian function $-\partial f_0/\partial v_z$, so that

$$\frac{\partial F}{\partial v_z} = \frac{D_c}{D_{hf}+D_c} \frac{\partial f_0}{\partial v_z}.$$ (2.148)

Since D_{hf} has a sharp maximum at $v_z = v_{ph}$ given by

$$D_{hf}\Big|_{v_z=v_{ph}} \simeq e^2 |E|^2 / [\ m_e^2 (v_c \omega^2)^{1/3}],$$

the condition that the nonlinear distortions in the distribution function obey $D_c \gg D_{hf}$ is of the same order as the condition (2.143).

When the opposite inequality is satisfied, the distribution function develops a plateau through a mechanism in which, as before, electron trapping in the potential wells of the wave plays an important part. A rigorous analysis [54] shows, however, that, as before, the relative contribution of collisions will be determined by the parameter D_c/D_{hf}. Thus, the damping decrement γ of the wave for $D_c/(D_{hf})_{max} \ll 1$ will be given by

$$\gamma = \gamma_L (D_c/D_{hf})^{3/4} K,$$ (2.149)

where γ_L is the linear Landau damping decrement and K is a factor that is close to unity.

2.12. QUASILINEAR THEORY OF WAVE DAMPING

We now consider the distortions in the distribution functions produced during absorption of nonmonochromatic waves. As in Section 2.11, we shall examine the one-dimensional absorption of Langmuir waves, representing both the wave field and the oscillating portion of the electron distribution function f_1 as Fourier expansions:

$$\left.\begin{aligned} E(z, t) &= \sum_k \mathrm{Re}\left\{E_k \exp\left[i(kz - \omega t)\right]\right\} \\ f_1(v, z, t) &= \sum_k \mathrm{Re}\left\{f_{1k} \exp\left[i(kz - \omega t)\right]\right\} \end{aligned}\right\}.$$ (2.150)

We shall assume that f_1 satisfies the linearized Eq. (2.129), and the averaged part of the distribution function F satisfies Eq. (2.132).

Equation (2.129) is easily solved separately for each Fourier harmonic. Then the collisional operator can be approximated by Eq. (2.133). When the nonlinear term in Eq. (2.132) is averaged, the terms with different k vanish and we again obtain an equation of the form (2.134) with the sole difference that in the expression for D_{hf} the summation is taken over all k in the wave. Since the approximation $v_{eff}[(\omega - kv_z)^2 + v_{eff}^2]^{-1} = \pi\delta(\omega - kv_z)$ can be used in the case of interest to us, we find

$$D_{\text{hf}} = \frac{\pi e^2}{2m_e^2} \sum_k E_k E_k^* \, \delta \, (\omega - k v_z).$$

(2.151)

In the case of a low collision frequency it is possible to neglect collisions entirely in Eq. (2.129). Then, however, in accordance with the Landau rule it is necessary to assume that the wave frequency ω has a positive imaginary part ω''. Then in the limit $\omega'' \to 0$ we again obtain Eq. (2.151). The result is thus independent of the specific form of the collision operator, so we can write

$$\frac{\partial F}{\partial t} + v_z \frac{\partial F}{\partial z} - \frac{\partial}{\partial v_z} D_{\text{hf}} \frac{\partial F}{\partial v_z} = \frac{\delta F}{\delta t},$$

(2.152)

where D_{hf} is given by Eq. (2.151) or its obvious integral analog for a continuous distribution in k obtained with Fourier integrals.

Equations of the type (2.152) are widely used in the theory of turbulent plasmas for describing the quasilinear relaxation of oscillations. Strictly speaking, these equations are justified only if the oscillations are random [i.e., the amplitudes E_k in the fields (2.150) must be regarded as random functions]. Nevertheless, they are often used in problems on rf heating where the random field requirement is usually not met. Then it is implicitly assumed that under real conditions the motion of the charged particles caused by the nonuniformity of the magnetic field leads to an effective averaging which is equivalent to averaging over a random distribution of fields. Thus, an electron which moves around the outer edge of the torus interacts repeatedly with the field at a given point and, because of the incommensurability of the period of the oscillations and the electron's time of flight, this interaction can be regarded as essentially random.

It is easy to understand that the limits of applicability of the quasilinear equations for the wave amplitudes are much wider for a finite k spectrum than for a monochromatic wave. Thus, if nonlinear trapping of electrons in the potential well of the individual partial waves is to be neglected, it is necessary that

$$\Delta \omega / k \gg \sqrt{|e \Phi_0| / m_e},$$

(2.153)

where $\Delta \omega / k$ is the interval of phase velocities for the waves under consideration and Φ is the order of magnitude of their potential amplitude. It is clear that for an excited wave spectrum that is not too narrow, this condition is considerably milder than Eq. (2.138) or (2.143). In particular, an equation of the form (2.148) is also applicable to the case $D_{\text{hf}} \gg D_c$ when a plateau develops in the distribution function over the entire range of velocities corresponding to the range of phase velocities of the excited waves. The physical mechanism for formation of the plateau, unlike in the case of a monochromatic wave, is the "diffusion" of resonant electrons on "partial" waves with different phase velocities, rather than nonlinear capture in the potential wells of the waves.

Equation (2.152) can be used directly to examine the formation of the electron distribution function during lower hybrid heating by a beam of slow waves of the the type specified by Eqs. (2.46) and (2.104) with a given spectral distribution over $N_\parallel$. As in the linear approximation, when $k_\perp^2 v_{Te}^2 / \omega_{Be}^2 \ll 1$ the thermal motion of the electrons perpendicular to the magnetic field has little effect on the propagation of these waves. It is also clear that the resonant wave–particle interaction primarily changes the component of the electron velocity parallel to the magnetic field. These circumstances make it possible to use Eq. (2.152) with a one-dimensional (in velocity space) quasilinear operator. Strictly speaking, a correct description of the anisotropy in $F(\mathbf{v})$ requires that the collisional operator in Eq. (2.152) also include the derivatives with respect to $v_\perp$ and the dependence of D_c on $v_\perp$. This does not, however, lead to a change in the qualitative picture of the phenomenon and estimates can be made using the one-dimensional formulas for $F(v_z)$ obtained in Section 2.11 by substituting D_{hf} from Eq. (2.151) into them. In the first approximation for the steady state it can be assumed that on every magnetic surface there is a local balance [in accordance with Eq. (2.147)] between quasilinear diffusion, which tends to form a plateau, and collisions, which Maxwellianize the distribution function. Then the distribution of lower hybrid waves in the plasma layer will be given, as before, by Eq. (2.104) where the imaginary part of ϵ_{zz} is calculated using the $F(v_z)$ obtained by solving Eq. (2.147). As Eq. (2.148) shows, quasilinear effects increase the damping length for each spectral harmonic by roughly a factor of $[(D_{\text{hf}} + D_c)/D_c]_{v_z=c/N_\parallel}$.

The evolution of the distribution function at a cyclotron resonance obeys a quasilinear equation similar to, but considerably more cumbersome than, Eq. (2.152) [29]. The increased complexity is related to a whole series of factors. First, the cyclotron acceleration of particles depends strongly on the polarization of the field, so changes generally occur in both the transverse and longitudinal particle velocities. Thus, the quasilinear diffusion operator in velocity space is no longer one dimensional but has a tensor diffusion coefficient which depends in a very complicated fashion on the field polarization, $N_\perp$, $N_\parallel$, and the cyclotron harmonic number s. We shall not give that complicated formula here, but will examine a few simple examples as illustrations.

For cyclotron heating of plasmas in toroidal systems, waves propagating at an angle to the magnetic field are of greatest interest. The following somewhat simplified formula has been proposed [29, 44] as an approximation for electron cyclotron absorption of extraordinary waves propagating at angles $N \mid \pi/2 - \theta \mid \gg v_{Te}/c$:

$$\frac{\partial F}{\partial t} = \frac{\pi e^2}{4m_e^2} \int dk \, \frac{\partial}{v_\perp \partial v_\perp} \left[v_\perp \mid E_{\bar{k}}^- \mid^2 \frac{\partial F}{\partial v_\perp} \right] \delta\left(\omega - \mid \omega_{Be} \mid - kv_z\right), \qquad (2.154)$$

where $E_{\bar{k}}^-$ is the spectral component of the wave electric field rotating in the direction of the cyclotron orbits of the electrons. In this equation the diffusion of resonant electrons in their longitudinal velocities (which is negligible under these conditions) and the contribution of the field component E_z to wave absorption have been neglected.

Equation (2.154) is very simple for a wave packet propagating at a single, fixed angle θ_0 to the magnetic field. Assuming that the spectral distribution of the field amplitude $\mid E_k \mid$ has the form

$$\left.\begin{array}{l} \mid E_k \mid^2 = \dfrac{E^2}{\pi (k_2 - k_1)} \, \delta\left(k_\perp^2 - k_z^2 \, \text{tg}^2 \, \theta_0\right) \\[2mm] \text{for } k_1 < k_z < k_2, \\[2mm] \mid E_k \mid^2 = 0 \quad \text{for } k_z < k_1, \, k_z > k_2, \end{array}\right\} \qquad (2.155)$$

and taking the integral with respect to **k** by using the δ-function, it is possible to obtain [29, 44]

$$\frac{\partial F}{\partial t} = \frac{\pi e^2}{4m_e^2} \frac{E^2}{\Delta\omega} \frac{1}{v_\perp} \frac{\partial}{\partial v_\perp} \left(v_\perp \frac{\partial F}{\partial v_\perp} \right), \qquad (2.156)$$

where $\Delta\omega = (\partial\omega/\partial k_z)(k_2 - k_1)$ is the width of the frequency spectrum. The solution of Eq. (2.156) is a Maxwellian function $F \propto T_\perp^{-1} \exp\left[-m_e v_\perp^2/2T_\perp(t)\right]$ with an effective temperature

$$T_\perp(t) = T_0(1 + t/t_0), \qquad (2.157)$$

where

$$1/t_0 = \pi e^2 E^2/(2m_e \Delta\omega T_0) = \pi \omega_{Be}^2 c^2 E^2/(\Delta\omega v_{Te}^2 B_0^2). \qquad (2.158)$$

We thus conclude that, subject to the neglect of collisions, this wave packet causes a linear growth in the perpendicular electron temperature over the entire range of parallel velocities that satisfy the resonance condition:

$$(\omega - |\omega_{Be}|)^2/k_2^2 < v_z^2 < (\omega - |\omega_{Be}|)/k_1^2. \qquad (2.159)$$

Evidently, collisions can Maxwellianize this distribution if

$$v_c t_0 \gg 1. \qquad (2.160)$$

In the above example, quasilinear effects at the cyclotron resonance were treated in a uniform magnetic field model. In real systems the field inhomogeneity changes the cyclotron interaction considerably. Because of the inhomogeneity of the magnetic field in toroidal systems, the condition (2.127) for cyclotron resonance of a particle with a wave is not satisfied in the entire plasma volume, but only at some surface and, unlike with a uniform magnetic field, a resonance can be realized for particles with practically any velocity v_z. Furthermore, as they move along their drift trajectories charged particles repeatedly intersect the resonance surface and, so to say, perform a doubly periodic motion at the cyclotron frequency ω_B and at the bounce frequency ω_{bd} along the drift trajectory. This causes "splitting" of the resonances [51], so that the resonance condition (2.127) will be replaced by

$$\omega = n\omega_B + l\omega_{bd} + k_{\parallel} v_{\parallel}, \qquad (2.161)$$

where n and l are integers and the subscript ∥ denotes the components of the velocity and wave vector along the total magnetic field.

Separate resonances of the form (2.161), however, obviously cannot appear under real conditions, primarily because during the time a resonant particle completes its drift orbit the phase between a wave and the particle may change substantially owing to collisions. In addition, "nonlinear randomization" processes may play an important role (see Section 2.13).

In simplified studies of ion cyclotron heating a model proposed and justified by Stix [59] is often used. This model makes formal use of an equation of the type (2.154) with the cyclotron frequency now regarded as a function of the coordinates, $\omega_{Bi} = \overline{\omega}_{Bi}[1 + (r/R)\cos\vartheta]$, for a monochromatic wave. This equation is averaged over the magnetic surface with the result that [51]

$$\frac{\partial \Gamma}{\partial t} = \frac{1}{v_\perp} \frac{\partial}{\partial v_\perp} v_\perp D_{hf} \frac{\partial F}{\partial v_\perp} + \frac{\delta F}{\delta t}, \qquad (2.162)$$

where

$$D_{hf} = \frac{e_i^2\, n}{m_i^2\,(2^n n!)^2} \left(\frac{k_\perp v_\perp}{\omega_{Bi}}\right)^{2n-2} \frac{|E_r + i E_\vartheta|^2 R}{\omega_{Bi}\,|\sin\vartheta_0|\, r}, \quad \omega = n\overline{\omega}_{Bi}\left(1 + \frac{r}{R}\cos\vartheta_0\right).$$

It is easy to see that, as in the case of the electron cyclotron resonance, when the condition

$$D_{hf}/v_{Ti}^2 \ll v_i \qquad (2.163)$$

is satisfied, where v_i^{-1} is the characteristic relaxation time for F, the bulk ion distribution is close to Maxwellian. Even when Eq. (2.163) is satisfied, however, the distribution function may differ from Maxwellian at high energies ($v_\perp^2 \gg v_T^2$). This happens because the quasilinear diffusion time $\tau_{QL} \propto v_{Ti}^4 (D_{hf} v_\perp^2)^{-1}$ decreases with increasing $v_\perp$ as $v_\perp^{-2n}$, while the Coulomb relaxation time τ_c increases rapidly, as $\tau_c \propto (1/v_i)(v_\perp/v_{Ti})^3$. Hence, quasilinear diffusion of fast ions predominates over Coulomb relaxation and high-energy "tails" are formed in the ion distribution.

2.13. STOCHASTIC ION HEATING IN THE LOWER HYBRID RESONANCE FREQUENCY RANGE

It follows from the dispersion equation (2.104) that the perpendicular refractive index $N_\perp$ of the slow lower hybrid wave can become extremely large near the conversion region $|\epsilon| \ll 1$ with the wavelength becoming much smaller than the ion Larmor radius. But, since the wave frequency is much higher than the ion cyclotron frequency, one might suppose that the interaction of an ion with the field of such a wave would differ little from the case of an unmagnetized plasma in which the unperturbed ion trajectories are straight lines. The ions, moving at a velocity close to the phase velocity of the wave, could interact resonantly with the wave for a long time through Landau damping and thereby dampen the wave.

It is clear from Section 2.6 that this kind of absorption does not occur in the linear approximation. For arbitrary $N_\perp$, wave damping occurs only through resonances at harmonics of the cyclotron frequency (2.127) and no resonance of the type $\omega = k_\perp v_\perp$ occurs for perpendicular motion. When nonlinear perturbations of the ion trajectories (on their cyclotron orbits) by the wave are taken into account, however, this type of absorption is possible [57, 58], although it differs somewhat from absorption by Landau damping in an unmagnetized plasma. In order to understand the physical mechanism for this absorption, we now consider the motion of the ions in the field of a longitudinal wave of the form

$$E_x = E \cos(kx - \omega t); \quad E_y = E_z = 0, \tag{2.164}$$

propagating perpendicular to an external magnetic field $\mathbf{B}_0$, directed, as usual, along the Z axis. We shall assume that the wave frequency ω and wave vector $\mathbf{k}$ satisfy the conditions

$$\omega \gg \omega_{Bi}, \quad k v_\perp \gg \omega_{Bi}, \tag{2.165}$$

where $v_\perp = \sqrt{v_x^2 + v_y^2}$. As noted in Section 2.8, the slow wave (2.104) is almost longitudinal with $N_\perp \gg N_z$, so that the field (2.164) is a reasonably satisfactory model of the actual situation.

The equations of motion of the ion in this field have the form

$$\dot{v}_x = \omega_{Bi} v_y + (e_i/m_i) E \cos(kx - \omega t), \tag{2.166}$$

$$\dot{v}_y = -\omega_{Bi} v_x, \tag{2.167}$$

where a dot denotes the time derivative.

It is convenient to introduce the new variable $v_x + i v_y = \xi \exp(-i\omega_{Bi}t)$. Multiplying Eq. (2.167) by i and adding it to Eq. (2.166), we obtain

$$\dot{\xi} = \frac{e_i E}{2m_i} \left\{ \exp(i[kx - (\omega - \omega_{Bi})t]) + \exp(-i[kx - (\omega + \omega_{Bi})t]) \right\}. \tag{2.168}$$

The coordinate x in this equation can easily be related to ξ through Eq. (2.167). Integrating this equation, we find

$$x = x_0 - \frac{v_y}{\omega_{Bi}} = x_0 + \frac{i}{2\omega_{Bi}} [\xi \exp(-i\omega_{Bi}t) - \xi^* \exp(i\omega_{Bi}t)], \tag{2.169}$$

where x_0 is a constant determined by the initial conditions. For the unperturbed motion ($E = 0$) it gives the location of the center of the cyclotron orbit of an ion. Then, of course, ξ is constant and equal to

$$\xi = v_\perp \exp(i\varphi), \tag{2.170}$$

where $\varphi = \varphi_0$ is the initial phase of the cyclotron orbit of the ion, which determines the initial ratio of v_x to v_y and the location of the ion in its orbit.

It is easy to see that the "effective force" on the right-hand side of Eq. (2.168) oscillates rapidly. Thus, it only produces a small, rapid oscillation in ξ. An exception occurs only at times when the phases of the terms on the right-hand side of Eq. (2.168) are stationary, i.e., at those times t_r when the time derivatives of the phases are zero:

$$k v_x(t_r) = \omega \pm \omega_{Bi}. \tag{2.171}$$

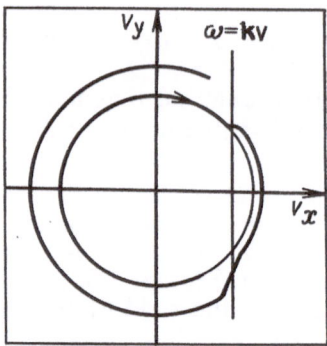

Fig. 2.19. Illustrating stochastic ion heating.

Here the $-$ and $+$ signs correspond to the first and second terms, respectively, on the right of Eq. (2.168). Because of condition (2.165), Eq. (2.171) is almost the same as the condition for the Landau resonance,

$$\omega = k v_x(t_r),\qquad(2.172)$$

which can only be satisfied for particles with $|\xi| = v_\perp \geq v_{ph} = \omega/k$. When $v_\perp > v_{ph}$ it is satisfied at two points on the cyclotron orbit (Fig. 2.19). The neighborhoods of these points are responsible for most of the change in ξ. This change is determined by the duration of the "resonant" interaction during which the phase is unable to change significantly. This time is related to the second derivative of the phase, $k\dot{v}_x$. Using Eq. (2.166) and assuming that the field E is fairly small, we can write

$$k\dot{v}_x \simeq k\,\omega_{Bi}v_y.\qquad(2.173)$$

The changes in ξ are greatest near points at which the first and second derivatives of the phase both go simultaneously to zero. We shall first examine these distinctive points, assuming that at time t_r the ion velocity component v_x satisfies Eq. (2.172) and $v_y = 0$. Neglecting for simplicity the term $\omega_{Bi}(t - t_r)$, which varies little during the time of acceleration, we approximate the phases of the terms on the right of Eq. (2.168) by

$$kx - (\omega \mp \omega_{Bi})\,t \simeq kx_r - (\omega \mp \omega_{Bi})\,t_r + \frac{k}{3!}\dddot{x}\,(t_r)(t - t_r)^3 + \ldots,\qquad(2.174)$$

where $x_r = x|_{t=t_r} = x_0$. In order to calculate $\dddot{x}(t_r)$ we use Eqs. (2.166), (2.167), and (2.172):

$$\dddot{x}\,(t_r) = -\omega_{Bi}^2\,\omega/k.\qquad(2.175)$$

By evaluating the fourth derivative of the phase it is easy to show that Eqs. (2.174) and (2.175) provide a sufficiently accurate estimate of the phase over a time much longer than the resonant interaction time. Thus, substituting this expression in Eq. (2.168) and integrating it over a fairly long time interval from $t_r - \tau$ to $t_r + \tau$, we find the increase in ξ during a single pass through the "resonance" zone by an ion in its orbit. The problem thus reduces to calculating the integrals

$$\int_{t_r-\tau}^{t_r+\tau} \exp\left[\mp\frac{i}{6}(\dot{\omega}_{Bi}^2\,\omega)(t-t_r)^3\right]dt.$$

Because of the rapid oscillations in the integrand, the main contribution to these integrals is from the range $(1/6)(\omega_{Bi}^2\omega)(t-t_r)^2 < 1$, so that the limits of integration can be extended to infinity. Then, by introducing the new variable of integration $\zeta = (\omega_{Bi}^2\omega/6)^{1/3}(t-t_r)$, they can be reduced to the tabulated form

$$\int_{-\infty}^{\infty} \exp(\mp i\zeta^3)d\zeta = (1/\sqrt{3})\,\Gamma(1/3),$$

where $\Gamma(1/3)$ is the gamma function. As a result, we obtain

$$\Delta\xi^{(1)} = v_E\exp(i\omega_{Bi}t_r)\cos(kx_r - \omega t_r), \qquad (2.176)$$

where $\Delta\xi^{(1)}$ is the growth in ξ during a single pass (at the time $t_r = 0$) of the "resonance" zone and $v_E = [e_iE\Gamma(1/3)/(\sqrt{3}\,m_i)](6/\omega_{Bi}^2\omega)^{1/3}$.

We now follow the process whereby an ion is accelerated during successive passes along its orbit. We begin with the linear approximation, where the field E is so small that the variation in x in the exponent of Eq. (2.168) owing to the growth in ξ [cf. Eq. (2.169)] can be neglected, and use Eq. (2.170) for the unperturbed orbit. It is easy to see that in this approximation the ion resonates with the wave at successive times $t_r^{(p)} = 2\pi(p-1)/\omega_{Bi}$, where p is the "number" of the passage along the orbit, and that after each such pass ξ changes by an amount given by Eq. (2.176) with $t_r = t_r^{(p)}$.

Using the formulas for a geometric progression, the total change in ξ after p passes can be written as

$$\sum_{s=1}^{p}\Delta\xi^{(s)} = v_E\frac{\sin(kx_0 + \pi\omega/\omega_{Bi}) - \sin[kx_0 - (\pi\omega/\omega_{Bi})(2p-1)]}{2\sin(\pi\omega/\omega_{Bi})}. \qquad (2.177)$$

Let us examine this formula near a cyclotron harmonic $\omega/\omega_{Bi} = n + \delta$, where $|\delta| \ll 1$ and n is an integer. In this case

$$\sum_{s=1}^{p}\Delta\xi^{(s)} \simeq \frac{1}{2\sqrt{3}}\left(\frac{6}{\omega_{Bi}^2\,\omega}\right)^{1/3}\frac{e_iE\Gamma(1/3)}{\pi m_i\,\delta}(1)^n \qquad (2.178)$$

$$\times\left\{\sin[kx_0 - (\omega - n\omega_{Bi})t_r^{(p)}] - \sin(kx_0)\right\}$$

and we conclude that ξ oscillates about its initial value with a frequency $\omega - n\omega_{Bi}$ and an amplitude proportional to $1/\delta = \omega_{Bi}/(\omega - n\omega_{Bi})$.

Equations (2.177) and (2.178) make it possible to understand physically why there is no linear Landau damping during propagation of a wave perpendicular to the magnetic field. Although a particle interacts resonantly with the wave during every pass on its cyclotron orbit, with incommensurate periods of the field and of the particle's cyclotron orbit, its energy does not change on the average over a long time interval. This is because when $\omega \neq n\omega_{Bi}$ the phase at which the particle enters the resonance region changes monotonically, so that the periods of its acceleration and deceleration in the field alternate regularly. When $\omega = n\omega_{Bi}$, this phase remains constant and a resonant change in the particle's energy occurs.

We now consider how the ion motion changes when the wave amplitude is large and the nonlinear perturbation of its trajectory cannot be neglected in Eq. (2.168). For a sufficiently low field the picture of an "elementary" interaction in a single passage of the "resonant" region is still valid and the velocity increase (growth in ξ) of the particle is again given by formulas similar to Eq. (2.176). Since the location of the resonance point x_r and the time from one resonance to another both vary when ξ changes, the phase of the particle as it enters the resonance also changes. When the phase of the particle as it enters the resonance increases noticeably relative to that of the wave (becomes comparable to $\pi/2$), the interaction of the ion with the wave changes qualitatively. The phases of successive entries of the ion into the resonance region become almost random and the energy accumulation process becomes diffusive [57, 58]. This phenomenon has come to be known as stochastic ion heating.

The change caused by the wave field in the phase with which the particles enter the resonance region, $\Delta(kx_r - \omega t_r)$, can be estimated using Eqs. (2.169), (2.172), and (2.176). When $|\cos kx_0| \approx 1$, it has the form

$$\Delta(kx_r - \omega t_r) \sim \frac{\omega}{\omega_{Bi}} \left(\frac{2kv_E}{\omega} \right)^{3/2} \sim \left(\frac{4kc}{\omega} \frac{E}{B_0} \right)^{3/2}.$$

The stochastic ion heating regime described above occurs when this phase change is comparable to unity, i.e., when

$$\frac{E}{B_0} > \frac{1}{4} \frac{\omega}{kc} \left(\frac{\omega_{Bi}}{\omega} \right)^{1/3}, \tag{2.179}$$

where the coefficient has been determined from an analysis of exact numerical solutions of Eqs. (2.166) and (2.167) [57].

The diffusive character of ion heating in the wave field makes it possible to use the Fokker–Planck equation for describing the evolution of their distribution functions [58]:

$$\frac{\partial F}{\partial t} = \frac{1}{v_\perp} \frac{\partial}{\partial v_\perp} v_\perp D_s \frac{\partial}{\partial v_\perp} F + \frac{\delta F}{\delta t}, \tag{2.180}$$

where D_s is the ion diffusion coefficient in velocity space and $\delta F/\delta t$ is the collision operator. The order of magnitude of the diffusion coefficient D_s when the threshold condition (2.179) is satisfied can be estimated as the ratio of the mean square of the growth in the perpendicular ion velocity in a single pass along its orbit to the cyclotron period. Karney [58] has analyzed Eq. (2.180) in detail with different model assumptions.

Chapter 3

ELECTRON CYCLOTRON HEATING

3.1. BASIC HEATING SCHEMES AND NUMERICAL SIMULATIONS

An approximate picture of wave propagation in toroidal plasmas has been given in Section 2.4 and linear wave absorption has been examined in Section 2.7. It follows from the considerations outlined there [5, 6, 44], that there are two possible schemes for heating plasmas in the electron cyclotron resonance range of frequencies. One is based on launching an ordinary wave from the outer edge of the torus (see Figs. 2.6 and 3.1). In order to ensure absorption of the electromagnetic energy at the center of the plasma in this case it is necessary, first, to choose a wave frequency such that the cyclotron resonance surface $| \omega_{Be}(r) | = \omega$ crosses the discharge axis or passes near it. Second, it is best to focus the beam of waves onto the center of the resonance surface (i.e., launch waves near the equatorial plane of the torus perpendicular to the toroidal field B_φ). As Eq. (2.89) shows, in large tokamaks of the size of T-15 and, even more so, under reactor conditions, the optical depth of the resonance layer for an ordinary wave propagating perpendicular to the magnetic field is so high that practically all of the wave energy is absorbed in a single "pass" through the resonance layer. Among the advantages of launching the wave in this way is the possibility of heating the plasma at the peak density $n_e \simeq n_c$ for the ordinary wave since (Section 2.4) an obliquely propagating wave is reflected earlier because of refraction. In addition, when a wave is launched perpendicular to the magnetic field B_0, the specifications for the antenna, which in this case excites an electric field E parallel to B_0, are simpler. When the ordinary wave is excited at an angle to B_0, the requirements for the polarization of the field excited by the antenna become much more complicated. As Eqs. (2.47) and (2.48) show, in general the field must be elliptically polarized with a certain direction of rotation of E.

In experiments on small tokamaks the optical depth of the cyclotron layer for the ordinary wave is often not enough for complete absorption and the remaining

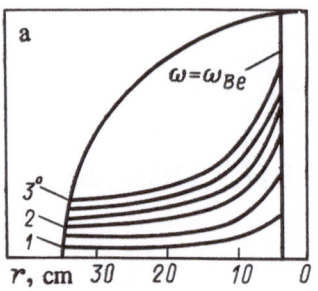 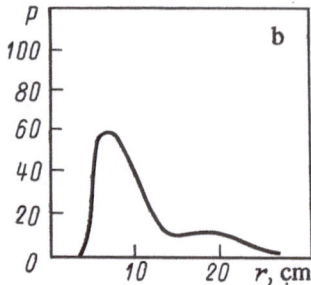

Fig. 3.1. Projections of ray trajectories of ordinary waves in the minor cross section of the T-10 vessel (a) and the energy deposition profile (b) [5].

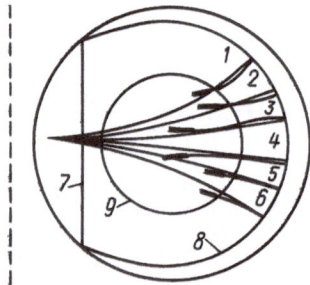

Fig. 3.2. Projections of ray trajectories of extraordinary waves in the minor cross section of T-15 [59]: The thick lines denote the portions of the trajectories on which 94% of the wave energy is absorbed; $\omega_p^2(0)/\omega^2 = 1.5$, $\omega_{Be}^2(0)/\omega^2 = 0.67$, and $T_{e0} = 5$ keV. The different curves correspond to $\delta = \arctan(k_{0r}/k_{0\varphi}) = 87°$ and to different values of $\alpha = \arctan(k_{0\theta}/k_{0r})$: 1) 7°; 2) 4.3°; 3) 1.4°; 4) −1.4°; 5) −4.3°; 6) −7°; 7) the electron cyclotron resonance curve; 8) the upper hybrid resonance curve; 9) the curve $r = 0.5a$.

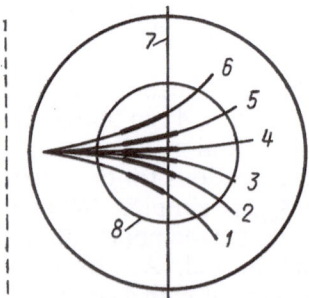

Fig. 3.3. Projections of ray trajectories of extraordinary waves in the minor cross section of T-15 [59] for $\omega_p^2(0)/\omega^2 = 1.6$, $\omega_{Be}^2(0)/\omega^2 = 1$, $T_e = 5$ keV, and $\delta = 73°$, with 1) $\alpha = -10°$; 2) −4.3°; 3) −1.4°; 4) 1.4°; 5) 4.3°; 6) 10°; 7) the electron cyclotron resonance curve; 8) the curve $r = 0.5a$.

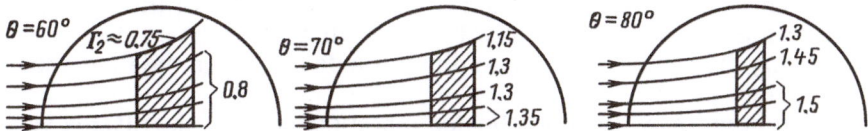

Fig. 3.4. Projections of ray trajectories of ordinary waves in the minor cross section of T-10 for different wave launch angles θ relative to the magnetic field; the region in which the microwave energy is deposited in the plasma is shaded; $\omega_{pe}^2(0)/\omega^2 \approx 0.5$, $\omega_{Be}^2(0)/\omega^2 = 1$ [60].

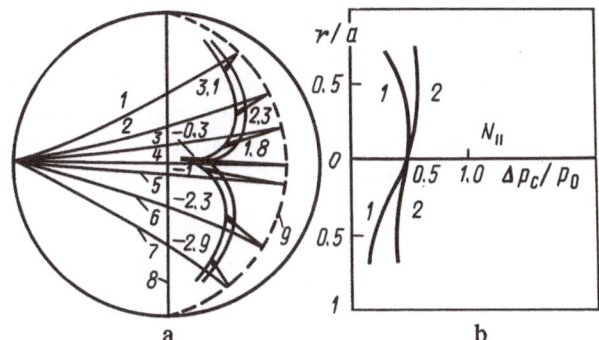

Fig. 3.5. (a) Projections of ray trajectories on the minor cross section of the FT-1 vessel for $\omega_{pe}^2(0)/\omega^2 = 0.6$, $\omega_{Be}^2(0)/\omega^2 = 1$, $\delta = 48°$: 1) $\alpha = 27°$; 2) 13°; 3) 6°; 4) –3°; 5) –6°; 6) –13°; 7) –27°; 8) the electron cyclotron resonance curve; 9) the upper hybrid resonance curve. (b) The dependence of $N_{\parallel}$ (2) and of the relative absorption of extraordinary wave power $\Delta P/P = 1 - \exp(-2\Gamma)$ in the electron cyclotron resonance region (1) on the location of the point of intersection of the electron cyclotron resonance curve with the ray trajectory.

part of the wave energy goes to the vessel walls. In this case the absorption depends strongly on the way the wave is reflected from the vessel wall and a more detailed analysis is needed to determine it.

The second scheme for electron cyclotron heating (ECH) is based on the use of the extraordinary wave launched from the inner edge of the torus (see Figs. 2.6, 3.2, and 3.3). In this case, Eq. (2.88) shows that an extraordinary wave propagating strictly perpendicular to $\mathbf{B}_0$ is usually weakly absorbed in the cyclotron layer and reaches the upper hybrid resonance surface. There it undergoes linear conversion to a Bernstein wave which is completely absorbed as it moves to the cyclotron layer. The absorption in the cyclotron layer of an extraordinary wave propagating at an angle to $\mathbf{B}_0$ increases rapidly with increasing $N_{\parallel}^2$ and is almost complete in large tokamaks for relatively small $N_{\parallel}^2$. The primary advantage of heating with the extraordinary mode is the ability to heat denser plasmas (up to $n_e \lesssim 2n_c$) than with the ordinary mode.

In principle it is possible to heat with an extraordinary wave at a higher frequency for which resonance at the second harmonic of the cyclotron frequency $\omega = 2|\omega_{Be}|$ occurs within the plasma. This wave is absorbed fairly efficiently in large tokamaks. The wave can be launched from the outer edge of the torus (see Fig. 2.7). The results of Section 2.4 show, however, that the limits on the plasma density are roughly the same as for extraordinary mode heating at the fundamental cyclotron resonance.

As noted previously, the complicated inhomogeneity of the plasma and magnetic field in tokamaks makes it difficult to perform a detailed analytic study of the heating behavior. Thus, numerical simulations have recently come into widespread use. Most of this work involves detailed numerical studies of the propagation and absorption of waves in tokamaks in the linear approximation, but with the actual geometry and plasma inhomogeneity taken into account (see, for example, the reviews by Alikaev et al. [5, 44], and the references cited therein). These studies are based on computing the ray trajectories and absorption of waves excited in the plasma using the approximation of geometric optics. Modelling the directional diagram of an antenna as a bundle of ray trajectories, with a suitable energy flux assigned to each, yields a plausible picture of localized energy deposition within a plasma when quasilinear distortions of the electron distribution function and other nonlinear effects are unimportant.

Considerable attention has been devoted to numerical simulation of the quasilinear distortions in the electron distribution function during ECH [44]. Generally, the plasma is treated as uniform in these simulations. At best, approximate models for inhomogeneous plasmas have been used along with the quasilinear diffusion coefficients from uniform plasma theory.

Finally, simulations of plasma heating have been done using transport codes in which the energy deposition profile was usually specified in a fairly simple form that was, nevertheless, in rough qualitative agreement with computational results on linear wave absorption [5].

We now discuss the principal results obtained from simulations of heating using the above schemes. Figures 3.1–3.6 illustrate the absorption of electromagnetic waves in tokamaks according to ray tracing calculations with realistic assumptions about the plasma inhomogeneity. These figures show the projections of ray trajectories in the minor cross section of tokamaks under various conditions and include data characterizing the efficiency and localization of deposition of the electromagnetic energy. Thus, Figs. 3.1a and 3.4 show a beam of ordinary mode ray trajectories in the minor cross section of a tokamak with the parameters of T-10 [5, 60] and Fig. 3.1b shows the distribution of the density of energy deposition over the cyclotron layer [5]. Figures 3.2 and 3.3 illustrate the ray trajectories of extraordinary waves in a tokamak the size of T-15 ($R_0 = 240$ cm, $a = 70$ cm, $I_p = 1.4$ MA, $T_{e0} \approx 5$ keV, $n_e(0) = 1.6 n_c$, $B_0 = 3.5$ T) [59]. In these figures a thick line on the trajectories denotes the parts on which 99% of the energy propagating along a given ray is absorbed. Figure 3.2 refers to the case where the extraordinary wave is weakly absorbed as it passes through the cyclotron layer and most of the electromagnetic energy is absorbed by damping of Bernstein waves formed through linear conversion in the upper hybrid resonance region. Figure 3.3 illustrates the case of oblique propagation of the extraordinary wave through the cyclotron layer when energy absorption is very efficient there. Figure 3.5 [61] shows the ray trajectories and absorption of extraordinary waves in a small tokamak the size of FT-1 ($R_0 = 62$ cm, $a = 15$ cm, $B_0 = 0.8–1.1$ T, $I_p \approx 30$ kA, $\bar{n}_e < 2 \cdot 10^{13}$ cm^{-3}). Here, even when the wave propagates at a noticeable angle to the magnetic field, cyclotron absorption of the wave is incomplete and linear conversion into a Bernstein wave plays an important role. The distribution of the poloidal magnetic field has a significant effect on the localization of cyclotron absorption of Bernstein waves. Indeed, under actual conditions the total magnetic field is nonuniform along the magnetic field lines because of the toroidal nonuniformity of the magnetic field. This clearly leads to changes in $N_{\parallel}$ along the ray trajectory which differ at different points in the minor cross section of the plasma. Since Bernstein waves are strongly damped as the condition $(\omega - |\omega_{Be}|) \simeq \omega N_{\parallel} v_{Te}/c$ is approached, these changes in $N_{\parallel}$ will evidently have a strong effect on the location where the energy of these waves is absorbed (Fig. 3.5).

As noted before, excitation of the extraordinary mode at an angle to $\mathbf{B}_0$ requires that the wave have a certain (elliptical) polarization at the plasma boundary [see Eqs. (2.47) and (2.48)]. It is difficult to obtain this polarization experimentally and it is usually obtained by superposing waves of both types. Then the ordinary wave is either reflected from the center of the plasma or, at low densities, passes through the resonance layer with little absorption and goes to the wall. In either case the subsequent fate of this wave is determined by the nature of the reflection from the vessel walls. Reflection of waves from the vessel walls is also important for a dense plasma with $n_e > 2n_c$ [61], when even the extraordinary wave cannot reach the center of the plasma (see Fig. 3.6). Since the vacuum vessels of tokamaks are usually bumpy (i.e., made of bellows sections), the waves are strongly scattered in the toroidal direction upon reflection and are depolarized. Thus, for the calculations in Fig. 3.6 [61] it was assumed that during each reflection the polarization of the wave and the toroidal projection of the wave vector change randomly. By approximating the directional diagram of the antenna with a sufficiently large number of rays, each of which has an "assigned" energy flux, it is possible to compute the energy deposition profile in the minor cross section of the plasma. These calculations have been done for the FT-1 tokamak [62]. They show that an increase in the plasma density shifts the absorption maximum toward the plasma edge, although this occurs much more slowly than might be expected from qualitative considerations.

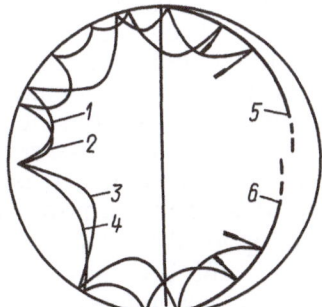

Fig. 3.6. Projections of ray trajectories in the minor cross section of FT-1: 1) $\omega_{pe}^2(0)/\omega^2 = 3$, $\alpha = 21°$, $\delta = 48°$; 2) $\omega_{pe}^2(0)/\omega^2 = 3$, $\alpha = 30°$, $\delta = 48°$; 3) $\omega_{pe}^2(0)/\omega^2 = 2$, $\alpha = -21°$, $\delta = 48°$; 4) $\omega_{pe}^2(0)/\omega^2 = 3$, $\alpha = -35°$, $\delta = -35°$; 5) the upper hybrid resonance curve for $\omega_{pe}^2(0)/\omega^2 = 3$; 6) the upper hybrid resonance curve for $\omega_{pe}^2(0)/\omega^2 = 2$.

A comparison of Figs. 3.1–3.6 clarifies the dependence of the propagation and absorption behavior of waves on their polarization and initial angles of incidence, on the density and temperature of the plasma, on the location of the resonance surface, and on the dimensions of the device. Clearly, with one or the other scheme for ECH it can be applied to machines of various sizes.

The basic features of the quasilinear evolution of the electron distribution function during ECH have been discussed briefly in Section 2.12. Using the criterion (2.160) given there, it can be shown that under typical experimental conditions the quasilinear effects do not cause a significant reduction in heating. (This question has been examined in more detail by Alikaev et al. [44].)

3.2. HEATING TECHNIQUES

For the magnetic fields in quasistationary systems the frequencies of the electron cyclotron resonance correspond to millimeter wavelengths: from 1 cm (at $B_0 \simeq$ 1 T) to 1 mm (at $B_0 \simeq$ 10 T). First of all, the realization of ECH requires high-power sources in this range. The power of traditional electronic microwave sources in the centimeter range (klystrons, magnetrons) falls off rapidly toward the millimeter wavelength range owing to a reduction in the size of the microwave systems in these sources (because of the reduced wavelength) and, therefore, to a reduction in the size of the cathodes. Hence, the most difficult technological problem standing in the way of using ECH in thermonuclear experiments is the creation of sufficiently powerful millimeter wavelength sources. At the present time, the gyrotrons developed at the Institute of Applied Physics of the Academy of Sciences of the USSR in Gorky [63, 64] appear to be suitable. In gyrotrons microwave power is generated by a relativistic electron beam moving along a helical trajectory in a magnetic field. The beam interacts with the microwave field through an open cylindrical cavity which ensures excitation of a given mode. Efficient generation occurs at the cyclotron resonance frequency or its second harmonic. Recently the Western companies Varian and Thomson [65, 66] have also undertaken the development of gyrotrons. Existing sources produce powers of 100–500 kW at wavelengths of from 3.6 to 10 mm in pulses lasting up to 1 s. Sources for shorter wavelengths and higher powers are under development.

Microwave power is usually channeled from the source by means of oversized multimode waveguides with cylindrical cross sections of diameter 2–8 cm. Because of the larger size of the waveguide, it is easy to prevent microwave breakdown in it. (In order to raise the peak power the waveguide is filled with compressed air to a pressure of up to 10^6 Pa.) Symmetric H_{01} modes propagate in uniform cylindrical waveguides with low loss, 10^{-4}–10^{-2} dB/m [67]. Because various modes can be excited at any nonuniformities in the waveguide, special steps must be taken. Turns, bends, and other waveguide elements must be made using quasioptical methods with filters for parasitic modes. In this way, transmission efficiencies of 70–90% to thermonuclear experiments over path lengths of 10–40 m can be obtained. One of the most difficult problems is diagnosing the mode composition of waves propagating in a multimode waveguide. This is done by means of a system of apertures in the waveguide with detectors mounted behind the apertures. The total energy transmitted through the transmission line is usually determined from the heating of a special absorber.

The system for supplying microwave energy to a tokamak or stellarator is important for an experiment. As mentioned previously, it must ensure the delivery of the ordinary mode from the outer edge of the torus or the delivery of the extraordinary mode from the inner edge. The ability to vary the angle between the direction of propagation of the wave and the magnetic field is also desirable, as well as the ability to focus the wave onto the electron cyclotron resonance region. The simplest input system involves the use of the TE_{11} mode in a cylindrical waveguide. This mode can be obtained from cylindrically symmetric TE_{01} modes by using standard microwave converters. The TE_{11} mode is almost linearly polarized in the center of the waveguide. Thus it can be delivered to the plasma simply through an aperture in the vessel with the aid of a waveguide or conical horn mounted on the end of the waveguide. Then a polarization close to that of the ordinary wave at normal incidence is obtained if the electric field of the wave is parallel to the toroidal magnetic field and a polarization close to that of the extraordinary wave is obtained if these fields are perpendicular. When the radiation is incident obliquely, in order to excite a "pure" mode (ordinary or extraordinary) the antenna must create a wave with elliptical polarization. Even with linear polarization, however, "contamination" from the parasitic mode is small when the direction is not far from normal. When the extraordinary mode is excited at angles of 90–45°, the ordinary mode "contamination" is less than 10% (in terms of power). Vacuum isolation of the waveguide is usually by means of a ceramic window mounted near the emitter, through which a power of up to 200 kW can then be delivered. In these systems, however, it is difficult to obtain a narrow directional diagram.

As an example Fig. 3.7 shows the antenna system used to deliver ordinary and extraordinary waves in the FT-1 tokamak. (The wavelength is about 1 cm.) Introducing a wave from the inner edge of the torus is usually difficult because the space inside the torus is occupied by the transformer and magnetic field coils. These difficulties were overcome by making the extraordinary mode enter FT-1 through a rectangular waveguide which "wound around" the plasma column in the shadow of the limiter (Fig. 3.7). The emitter was attached to the outlet of this waveguide on the inside edge of the torus. In order to prevent breakdown in the waveguide in the cyclotron resonance region (which intersects the waveguide as it winds around the vessel), a ceramic vacuum window was mounted near the emitter and the waveguide was filled with compressed air. A similar scheme for delivering microwaves to the inner edge of the plasma was used on the JFT-2 tokamak [69]. In this device the antenna was an eight-waveguide slit structure which produced an elliptical polarization corresponding to the extraordinary wave. Then, by varying the phase between the waveguides it was possible to change the direction of the radiation relative to the magnetic field.

An improved input system for controlling the direction, polarization, and focussing of the radiation can be constructed using quasioptical elements (mirrors and prisms). Such a system has been used on the PDX tokamak (Fig. 3.8). There the TE_{02} mode from a gyrotron was first converted to the TE_{11} mode and then into an

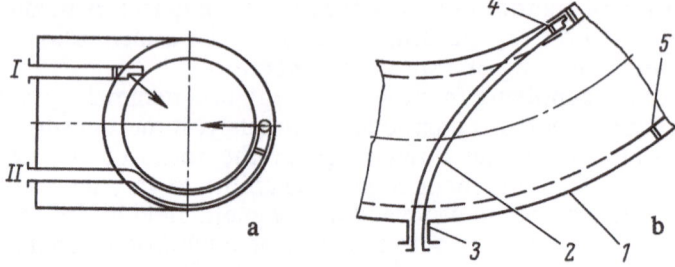

Fig. 3.7. The system for launching microwaves into the FT-1 tokamak plasma [68]: a) schemes for launching the ordinary (I) and extraordinary (II) waves; b) top view of the inner feedthrough [1) vessel; 2) waveguide; 3) feedthrough; 4) vacuum window; 5) limiter].

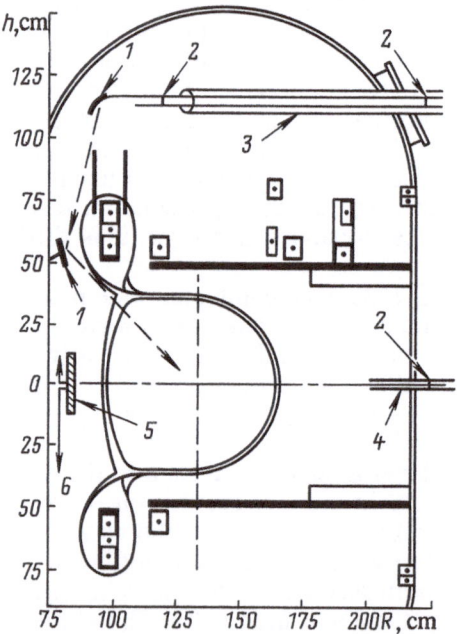

Fig. 3.8. The system for launching 5-mm microwaves into the PDX tokamak [71]: 1) mirror; 2) window; 3) inside launch; 4) outside launch; 5) polarized reflector; 6) to microwave measurement system.

HE_{11} mode with very nearly linear polarization. The efficiency of energy transmission through the roughly 40-m-long waveguide system and conversion into the required mode was 70–80%. The HE_{11} waveguide mode was close to quasioptical and emitted a thin beam with a nearly Gaussian transverse energy distribution [70]. Energy was fed in from the outside at an angle of 7° to the normal. The directional diagram corresponded to 10 dB attenuation of the field within an angle of 10° [71]. This made it possible to focus the radiation onto a volume with linear dimensions of about 10 cm, considerably smaller than the plasma radius (40 cm). One characteristic feature of this input system is the use of a reflecting mirror on the inner wall of the torus, mounted so that the ordinary wave polarization is converted through reflection into that of the extraordinary wave. This conversion means that an ordinary wave launched from the outside of the torus, which is not absorbed during its first passage through the plasma, will be absorbed after reflection from the inner wall as an extraordinary wave. The reflecting mirror offered an additional diagnostic possibility. It contained a number of holes through which the spatial distribution of the power passing through the plasma could be measured. In PDX, microwaves were also launched from the inside with the aid of two mirrors which created a quasioptical beam at an angle of 40° to the magnetic field and 45° to the horizontal plane. A similar system has been built on the W-VIIA stellarator for frequencies of 28 and 70 GHz [72]. More careful fabrication of the components in this system resulted in an energy transmission efficiency from the gyrotron to the antenna of about 90%.

The highest power ECH system in use is on the T-10 tokamak [73]. There the microwave power is generated by a special set of four gyrotrons operating at a wavelength of 3.6 mm with a power of up to 250 kW each and a pulse duration of up to 0.2 s. The magnetic field for the gyrotrons is produced by a cryogenic system. The waveguides that transmit the energy have diameters of 80 mm and are filled with SF_6 at a pressure of about 10^5 Pa to prevent breakdown. Power is delivered to the outside edge of the torus through circular waveguides in which the TE_{11} mode is excited in an orientation corresponding to emission of the ordinary mode. The efficiency of the input system is 70–80% with a path length of up to 10 m. The directional diagram of the antenna was measured in model experiments on a special stand and extended out to 8°. In the tokamak itself the wave propagation behavior can be determined using receiver antennas mounted at different places in the limiter shadow.

The use of ECH in a reactor will require a large increase in the power delivered to the plasma. The question of the optimum power for a single gyrotron module then arises. Estimates [74] indicate that reliable operation of a gyrotron with output pulse durations of 1–10 s can be assured at powers of up to 1 MW. Then the gyrotron system for heating a reactor plasma must consist of 100 gyrotrons. The energy from each gyrotron can probably be delivered without breakdown using a waveguide with a diameter of 5–10 cm and ceramic windows of the same size for vacuum isolation. The total area of these windows will be a fraction of a percent of the surface area of the vacuum vessel, so delivering the power in this way is quite suitable for a reactor. The main problem is to create reliable gyrotrons with the required power and wavelength, together with a reliable system for controlling their operation. An alternative gyrotron system could be based on combining the vacuum system of the gyrotrons with that of the tokamak [74]. Then a substantial increase in the power per gyrotron may be expected, but major difficulties will arise in connection with maintaining the vacuum.

3.3. EXPERIMENTAL RESULTS

The first experiments on ECH in tokamaks were done in the Soviet Union on the small Tuman-2, TM-3, and FT-1 machines [27, 75–77]. In these experiments

Table 3.1. Parameters of ECH Experiments on Tokamaks

Machine	Refs.	a (cm)	B (T)	$\bar{n}$ (10^{13} cm^{-3})	T_{e0} (keV)	f (GHz)	P (kW)	τ (ms)	Launch
T-10	[78,79]	35	1.5–3	2–6	1	83	1000	200	Outside
FT-1	[68,80]	15	1	0.3–2	0.5	30	150	3	Inside, outside
ISX-B	[81]	26	1.2	0.5–1	0.8	35	100	10	Inside
JFT-2	[82,83]	28	1	0.3–1.5	0.6	28	100	20	Inside, outside
Tosca	[84]	8.5	0.5	0.2–1	0.4	28	200	2	Outside
Doublet-III	[85]	44	2	1–6	1	60	800	80	Inside
PDX	[71]	44	2	1–2	1	60	100	100	Outside

Table 3.2. Parameters of ECH Experiments on Stellarators

Machine	Refs.	Machine parameters					Heating parameters		
		a (cm)	B (T)	I (kA)	$\bar{n}$ (10^{13} cm^{-3})	T_e (keV)	f (GHz)	P (kW)	τ (ms)
JIPP-TII	[86]	17	1.3	30	0.5	0.7	28	200	10
Heliotron E	[87]	20	1	0	0.5	0.5	28	200	20
Heliotron E	[88]	20	2	0	1–2	0.5	53	200	50
W-VII A	[89]	10	1	0	0.3–0.5	0.5–1	28	200	40
L-2	[90]	10	1	0	0.5	0.5	30	100	10

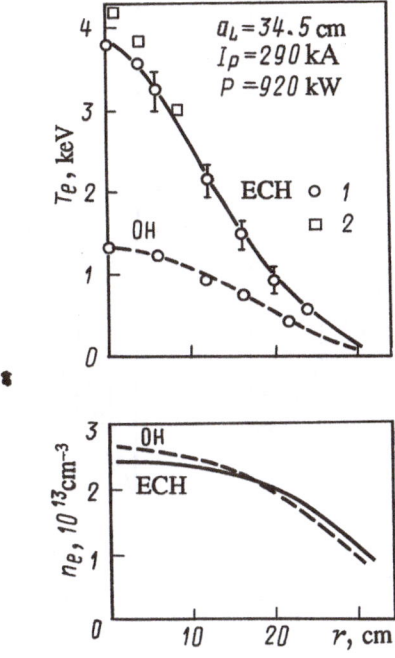

Fig. 3.9. Electron temperature and density profiles with ECH in T-10 [79] ($P = 920$ kW, $f = 83$ GHz): 1) soft x-ray; 2) cyclotron emission measurements of T_e.

the sources were gyrotrons with wavelengths of 0.8–1 and 0.4 cm, powers of up to 100 kW, and pulse durations of 1–3 ms. The microwave power was delivered to the vessel from the outside of the torus through the open end of a circular waveguide. Plasma heating was obtained at the electron cyclotron resonance frequency $\omega = |\omega_{Be}|$ and its second harmonic $\omega = 2|\omega_{Be}|$ in these experiments. It was found that heating was related to the location of the cyclotron resonance region and to fulfillment of the condition for accessibility of this region. The heating efficiency ρ_h for $\omega \approx |\omega_{Be}|$ reached 20–40%, a comparatively low value because the ordinary wave is weakly absorbed in a single pass for the plasma parameters typical of small tokamaks. On the other hand, the extraordinary wave introduced from the outside of the torus cannot penetrate into the resonance region and avoid the opacity region (see Fig. 2.6). Under these conditions the wave could undergo cyclotron absorption only as a result of several successive transits within the vacuum vessel after being reflected from the walls. Since each reflection leads to changes in the polarization and propagation direction of the wave, a substantial part of the wave energy was absorbed at the edge of the plasma in such an arrangement. This also caused a reduction in the heating efficiency.

A new series of experiments on ECH in tokamaks and stellarators was undertaken in 1979–1984. An important feature of most of these experiments was the use of devices for delivering the microwave energy with the required polarization and focussing (see Section 3.2). The parameters of these experiments are listed in Tables 3.1 and 3.2. In these experiments the ordinary wave was launched from the outer edge of the torus and the extraordinary wave, from the inner. In most ex-

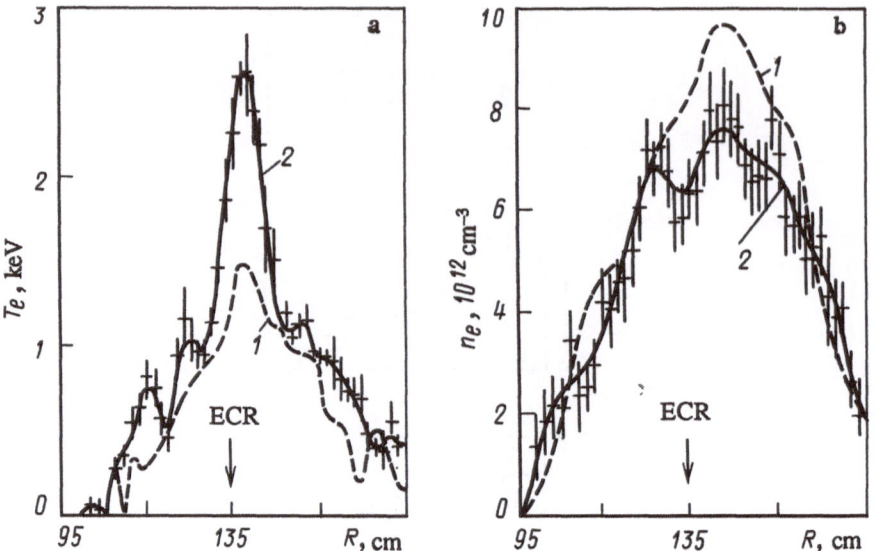

Fig. 3.10. Thomson scattering electron temperature (a) and density (b) profiles with ECH in PDX [71] ($P = 70$ kW, $f = 60$ GHz): 1) ohmic heating; 2) ECH. (ECR denotes the location of the electron cyclotron resonance.)

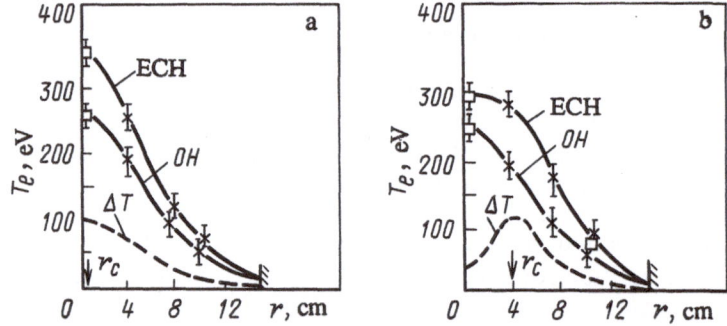

Fig. 3.11. Thomson scattering electron temperature profiles with ECH in FT-1 [62] ($P = 80$ kW, $f = 30$ GHz): (a) electron cyclotron resonance on axis; (b) electron cyclotron resonance shifted by 4 cm.

periments heating was at the electron cyclotron resonance frequency $\omega \sim |\omega_{Be}|$ and in some cases heating at the second harmonic $\omega \sim 2|\omega_{Be}|$ was studied as well.

The largest scale ECH experiment has been on the T-10 tokamak [78, 79]. Up to 1 MW of power at 83 GHz was delivered in 0.2-s pulses to the plasma by the gyrotron system described in Section 3.2. The ECH emitter was located on the outside edge of the torus and delivered 70% of the input power in the ordinary mode. The heating was most efficient (up to 80%) when the electron cyclotron resonance region was at the plasma center and the density was below critical ($n_e < n_c$). Figure 3.9 shows the change in the plasma parameters during ECH on T-10. The largest increase in T_e with $\bar{n} \approx 3 \cdot 10^{13}$ cm^{-3} and $P \sim 1$ MW was $\Delta T_e \sim 3$ keV, which corresponds to a normalized heating parameter $\eta_h \sim (8-10) \cdot 10^{13}$

eV/(kW·cm^3). The ion temperature increase was relatively small ($\Delta T_i \approx 0.2$ keV) and was caused by collisional energy transfer from the electrons to the ions. It should be noted that the microwave heating power on T-10 was considerably greater than the ohmic heating power. Before introduction of the microwaves the ohmic power was roughly 300 kW. The loop voltage was reduced by the ECH, so that the ohmic heating power fell to 100–200 kW. During ECH the radiation from impurities increased; however, the energy losses associated with this radiation on T-10, as in all other experiments on ECH, was still considerably lower than the heating power. Z_{eff} also changed little.

In T-10, as in the experiments on ISX-B and JFT-2, a drop in the average plasma density was observed during microwave heating. This drop was as much as 10–30%, with the greatest reduction in n_e at the plasma center. The reason for the drop in n_e is not yet clear. It may be caused by enhanced particle transport or by changes in recycling during microwave heating.

The localization of ECH means that it can be used to vary the electron temperature profile. This has been observed in a number of experiments. When the resonance zone is centrally located and the radiation is not very narrowly focussed, heating leads to an increase in the electron temperature without significant changes in the profile shape. One might suppose that under these conditions the distribution of heating power over the plasma cross section would be close to the ohmic heating power distribution. In an experiment on the PDX tokamak with "narrow" focussing of the radiation, strong peaking of the T_e profile was observed (Fig. 3.10). As the electron cyclotron resonance zone was shifted to the edge, of course, the radial T_e profile broadened. This is confirmed by measurements on the FT-1 and T-10 tokamaks (Fig. 3.11). A similar broadening of the T_e profile has also been observed during heating of plasmas with a peak density close to or greater than the cutoff density. In this case the broadening is caused by reflection and refraction of waves near the boundary of the opacity region, so that the absorption region is shifted toward the edge. (This is also suggested by numerical simulations of wave propagation, as in Fig. 3.1.) It is interesting to note that the outward shift of the cyclotron resonance region in low q discharges leads to suppression of sawtooth oscillations (or of the so-called internal disruption instability). This effect has been observed in T-10 and PDX [71, 79]. It is natural to relate this to broadening of the $T_e(r)$ profile. On the other hand, when $q(0) > 1$ with ohmic heating and there are no sawtooth oscillations, strong central heating may cause the appearance of oscillations associated with a drop in $q(0)$. This effect was observed during ECH on PDX.

It should be noted that ECH at low densities [$n_e < (0.5–1)\cdot10^{13}$ cm^{-3}] has a number of peculiarities. The ordinary wave is usually weakly absorbed in this regime. The extraordinary wave is almost completely absorbed (partly as a result of direct cyclotron damping and partly after conversion into the slow mode). Damping of the wave by fast electrons causes strong distortions in the distribution function since energy relaxation of the electrons is slow at low densities. In addition the resulting increase in the perpendicular energy leads to formation of locally trapped electrons. This manifests itself as strong suprathermal emission observed in many experiments at harmonics of the electron cyclotron frequency and in the soft x-ray range.

Unexpected fast ion production has been observed during ECH in the FT-1 tokamak [91]. Fast ions were detected through measurements of the spectrum of charge exchange neutrals emitted by the plasma. A sharp increase in the number of fast particles during ECH was accompanied by the appearance of a (parametric) decay spectrum in which one of the product waves was the lower hybrid wave. The threshold for this decay process was comparatively low ($P \sim 10$ kW). The decay apparently occurs in the neighborhood of the upper hybrid resonance zone. Lower hybrid waves formed during the decay process heat fast ions in resonances at harmonics of the ion cyclotron frequency or by stochastic heating (Section 2.13). Estimates show that only a small fraction (less than 5%) of the microwave power applied to the plasma is expended in ion heating by the waves formed in the decay process.

One of the central problems in ECH experiments has been to determine the heating efficiency. Information on the overall (integrated) efficiency is usually obtained from the ratio of the energy stored in the plasma electrons during ohmic heating alone to that when the ECH is turned on (Section 1.5). The plasma energy is most accurately determined from data on the electron density and temperature profiles; in some experiments diamagnetic measurements are made for this purpose. The heating efficiency seems to be greatest when the electron cyclotron resonance region is located in the plasma center, where refraction or reflection from the opacity region do not impede penetration of the wave into the resonance region and the optical depth of the resonance region is large enough ($\Gamma > 1$). For an extraordinary wave launched from the high field side this condition is realized for a peak density $n_{e0} < 2n_c(1 - N_{\parallel}^2)$. The optical depth of the electron cyclotron resonance region for an extraordinary wave launched at angle of 60–70° to the magnetic field is fairly large ($\Gamma \geq 1$) for the parameters of present-day tokamaks. When the extraordinary wave propagates nearly perpendicularly to the field, it is hardly absorbed in the resonance region. Then the wave reaches the upper hybrid resonance surface and is reflected from it after conversion into a slow Bernstein mode which should undergo strong cyclotron absorption (Section 3.1). Measurements of the dependence of central heating on the plasma density and magnetic field in the JFT-2 tokamak [83] show that in this case substantial absorption is localized near the hybrid resonance surface. This effect was attributed [83] to stochastic heating, although it may be also be related to the change in $N_{\parallel}$ discussed in Section 3.1. For the ordinary mode there is no transparency region in the center when $n_{e0} < n_c$. Then the optical depth of the electron cyclotron resonance region is high ($\Gamma > 1$) only in the very largest tokamaks. Under conditions such that the wave is absorbed in the plasma center, it is possible to obtain ECH efficiencies approaching those for ohmic heating ($\rho_h = 0.5$–1). It should be noted that fairly high heating efficiencies have also been obtained with heating at the second harmonic of the cyclotron frequency ($\omega \approx 2|\omega_{Be}|$) using the extraordinary wave. Then there is no opacity zone if $n_{e0} < n_c(\omega_{Be})$, and for the parameters of present-day tokamaks the optical depth of the resonance zone for the extraordinary wave, $\Gamma > 1$. Consequently, the efficiency ρ_h is on the order of 0.5.

Shifting the absorption zone toward the plasma edge as the electron cyclotron resonance surface is moved, increasing the plasma density, or changing to discharges with a low optical depth ($\Gamma < 1$) all lead to a reduction in the heating efficiency. According to the available data, the overall efficiency of absorption of microwave power is still large in all these cases. This is indicated by measurements of the reflected signal in a number of experiments, as well as of the rate of rise of the plasma energy during the initial stage of heating.

One of the problems associated with ECH originates in the need to raise the limiting plasma density at which heating is possible. The density dependence of the

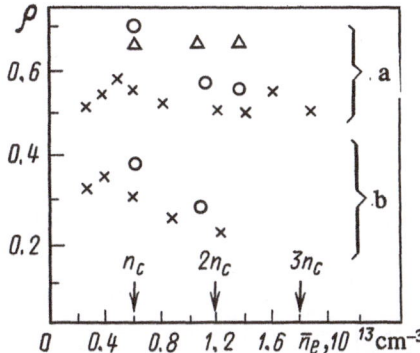

Fig. 3.12. The density dependence of the efficiency of ECH in FT-1 [68] for (a) inside launch of the extraordinary wave and (b) outside launch of the ordinary wave according to laser (Δ), diamagnetic ($\circ$), and conductivity ($\times$) measurements. The arrows denote values of n_{max}.

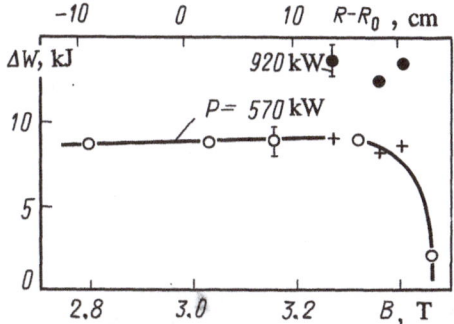

Fig. 3.13. The dependence of the energy increase from ECH in T-10 on the location of the electron cyclotron resonance region [79] ($n_{max} \sim n_c$).

heating efficiency has been studied on FT-1, JFT-2, and T-10 [62, 68, 79, 83]. The dependence of the increase in the central electron temperature, ΔT_{e0}, on the electron density has been determined for JFT-2. It was found that the ordinary mode heating efficiency at the center decreases for peak densities above critical, when an opacity region develops in the center of the plasma column and the absorption zone is shifted toward the edge. Thus, even when the extraordinary wave is launched from the inside of the torus, there is less heating in the center for densities approaching the cutoff density [$n_{e0} = 2n_c(1 - N_{\parallel}^2)$]. Experiments on FT-1 have shown, however, that the overall heating efficiency does not fall rapidly with increasing density. This is indicated by measurements of the efficiency as a function of the plasma density (Fig. 3.12). These data show that a comparatively high heating efficiency is maintained for the extraordinary wave even with peak densities $n_{e0} \simeq 3n_c$, when the opacity region is fairly large ($r \approx 0.3a$). A numerical simulation of this experiment showed [62] that this result can be explained by absorption during several successive reflections of the wave from the opacity region and from the

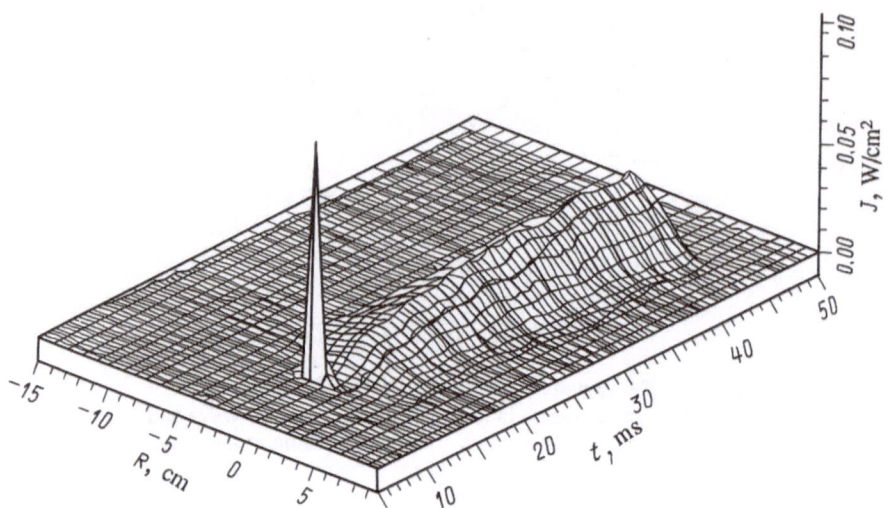

Fig. 3.14. The change in the soft x-ray emission profile during plasma formation by ECH in the W-VIIA stellarator [89].

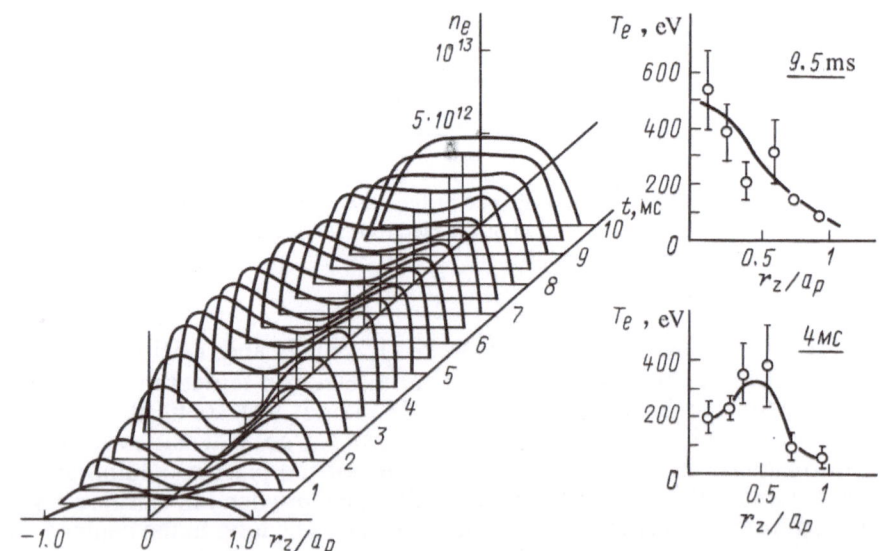

Fig. 3.15. Electron density and temperature profiles during ECH in the Heliotron E stellarator [87] ($P \approx 200$ kW; $f \approx 28$ GHz); the T_e profile was measured by Thomson scattering and the n_e profile with a microwave interferometer.

vessel walls including conversion of the extraordinary wave into a plasma wave in the upper hybrid resonance region. Although the main absorption zone is then located at fairly large radii $r = (0.6–0.8)a$, the overall heating efficiency is still high ($\rho_h = 0.4–0.5$).

In the T-10 tokamak with ordinary wave heating it was shown that a comparatively high efficiency is also maintained when the absorption region is shifted toward the edge (Fig. 3.13). This apparently happens because the effective thermal conductivity has a minimum in the edge region.

The influence of ECH on the energy confinement time is of considerable interest. It can be estimated from the steady-state energy balance and from data on the fall in the electron energy after the heating pulse ends (Section 1.5). In experiments with ECH powers comparable to or less than the ohmic heating power, no significant changes in τ_E are observed. With ECH power levels greater than the ohmic heating power in T-10 some reduction in the energy confinement time has been observed which depends on the conditions under which the vacuum vessel was outgassed. At the highest powers ($P \sim 1$ MW), τ_E was found to fall by a factor of 1.5–2 [79]. This effect was not regarded as specific to ECH, but presumed to be caused by a growth in the electron thermal conductivity with the electron temperature [79].

We now consider experiments on ECH in stellarators (see Table 3.2). The most interesting experiments have been conducted on the Heliotron E and W-VIIA stellarators [87–89]. In these experiments there was no ohmic heating phase and microwaves were used both to produce the plasma and to heat the electrons. A quantitative analysis of the conditions for wave absorption in stellarators is more cumbersome than in tokamaks because of the more complicated magnetic field geometry of the former. However, similar qualitative considerations apply to the conditions and mechanisms for heating in both.

Plasma production by microwaves occurs in the electron cyclotron resonance region. When this region is located inside the volume, plasma formation begins from within. This is shown especially by measurements of the radial distribution of soft x-ray radiation during the early stage of formation (Fig. 3.14). Further heating leads to increases in the electron density and temperature. The maximum density is limited by the criterion for wave propagation $n_{e0} < n_e$. In these experiments $n_{e0} \sim 10^{13}$ cm^{-3} for $\lambda \simeq 8$ mm and $n_{e0} \sim 4 \cdot 10^{13}$ cm^{-3} for $\lambda \simeq 4$ mm. The energy stored in the plasma depends roughly linearly on the power in the range studied ($P < 200$ kW). Figure 3.15 shows an example of the evolution of the T_e and n_e profiles during ECH in Heliotron E. The energy confinement time in the current-free plasma was determined from the steady-state energy balance and from the drop in the electron temperature after the microwave pulse. According to preliminary data, τ_E was considerably greater than the values predicted by scaling from ohmically heated tokamaks, although still lower than the predictions of neoclassical theory.

In conclusion, we note that ECH offers possibilities for investigating physical processes in toroidal plasmas since it provides a means independent of the current for increasing the electron temperature and controlling its radial profile. Even now, important physical problems in controlled thermonuclear fusion research are being solved with the aid of ECH, including: studies of the dependence of transport and MHD stability coefficients on the electron temperature and its radial profile, the production of high β in tokamaks and the finding of reasons for β limits, studies of the effect of anisotropies in the distribution function on the characteristics of tokamak discharges [79, 84], and the comparison of current-carrying and current-free discharges in stellarators [87, 89, 90].

3.4. SUMMARY OF RESEARCH

The successful development of high-power gyrotron sources of millimeter waves has led in recent years to widespread research on ECH. In experiments on tokamaks and stellarators two schemes for heating near the electron cyclotron resonance have been studied: using the extraordinary wave launched from the high magnetic field side and the using the ordinary wave launched from the low field side. Heating using the extraordinary wave near the second harmonic of the cyclotron frequency has also been studied. To excite these waves in plasmas, efficient waveguide and quasioptical emitters have been developed with emission power levels of up to 300 kW (power density up to 3 kW/cm^2). These experiments have yielded efficient local heating of the plasma electrons. Heating efficiencies comparable to that of ohmic heating, 50–90%, have been reached under various conditions. The possibility of controlling the electron temperature profile over fairly wide limits with ECH by shifting the location of the resonance has been demonstrated. It has been shown that the stability conditions in the plasma can be changed by this means. Some lowering of the plasma density (up to 30%) has been observed in heating experiments. At the same time, the additional influx of impurities was small and did not inhibit heating. The limitations on heating at high densities have been studied and it has been shown that the heating efficiency changes little up to the highest densities (close to the cutoff densities) and in some cases only decreases slowly at high densities. The highest powers have been applied to plasmas in the T-10 tokamak: up to 1 MW at a wavelength of 3.6 mm in pulses lasting 0.2 s. At this power level, electron temperature increases of $\Delta T_e = 3$ keV have been obtained at plasma densities on the order of $3 \cdot 10^{13}$ cm^{-3}, so that $\eta_h \approx 10^{14}$ eV/(kW·cm^3).

The main difficulty with using ECH in a thermonuclear reactor is making sources that have the required power, wavelength, and pulse duration (P up to 100 MW, $\lambda \approx 2$ mm, and $\tau \approx 10$ s). Such systems can, in principle, be based on gyrotrons with powers of 0.5–1 MW. The engineering development of such systems, however, is still in the future.

Limitations on the density and, accordingly, on the relative plasma pressure β are important considerations in the use of ECH. When the ordinary mode is used for heating, the attainable β is somewhat lower than required. Studies have shown that it can be raised by a factor of 2 by going to the extraordinary mode. This, however, involves a more complicated emitter design, which must be mounted on the inner edge of the torus. A considerable increase in the density and maximum β (by a factor of 4) can be achieved by going to ECH at the second harmonic using the ordinary mode. This approach, however, requires a doubling of the source frequency, so that it will be more difficult to realize.

Chapter 4

LOWER HYBRID HEATING

4.1. THEORETICAL MODEL

Rf heating of tokamak plasmas in the lower hybrid resonance range of frequencies, based on the absorption of strongly damped waves by the plasma electrons and ions (lower hybrid heating, LHH), is under extensive consideration. From an engineering standpoint the lower hybrid resonance range of frequencies, which falls in the decimeter wavelength range for the parameters of thermonuclear experiments, is extremely attractive. On one hand, in this range it is possible to use highly developed and efficient sources, klystrons or magnetrons. In this regard it differs favorably from the higher frequency range of the electron cyclotron resonance. On the other hand, at decimeter wavelengths waveguide structures can be used for exciting waves in plasmas. It is much easier to mount these structures in a tokamak than the ones used for the lower frequency ion cyclotron resonance range.

As in the other frequency ranges, a discussion of lower hybrid resonance plasma heating involves a range of topics, including:

the processes which control the input of rf energy into tokamak plasmas;
the propagation of waves from the plasma edge to the center;
the mechanisms for energy absorption and heating; and
the effect of heating on plasma confinement in the tokamak.

Here the propagation behavior of waves plays a fundamental role, substantially determining the specifications for the input structure and the mechanisms of energy absorption that lead to efficient heating. Accordingly, our discussion of current theoretical ideas about LHH begins with a simplified picture of wave propagation [92, 94] based on a one-dimensional inhomogeneous plasma model (see the review by Tonon et al. [93], as well). We then discuss various refinements in this picture (toroidal effects, nonlinear wave interactions, nonlinear heating mechanisms, etc.).

In the first approximation the basic propagation behavior of waves in a tokamak in the frequency range $\omega^2 < |\omega_{Be}|\omega_{Bi}$ can be understood by starting with a plane-layered, cold plasma model (Section 2.4) which, while far from the actual geometry of a tokamak, does take one of the main factors into account: the density rise from the edge toward the center of the plasma. In this model two geometric-optical waves which are slowed down along the external magnetic field can propagate. The approximate dispersion equations for these waves can be written as

$$\epsilon N_\perp^2 = \eta\,(\epsilon - N_\parallel^2) - g^2\,\eta N_\parallel^2/\,(\epsilon - N_\parallel^2)\,(\epsilon - \eta) \qquad (4.1)$$

and

$$(\epsilon - N_\parallel^2)\,N_\perp^2 = (\epsilon - N_\parallel^2)^2 - g^2, \qquad (4.2)$$

where $\epsilon = 1 + \omega_{pe}^2/\omega_{Be}^2 - \omega_{pi}^2/\omega^2$; $\eta = 1 - \omega_{pe}^2/\omega^2$; $g = \omega_{pe}^2/(\omega\omega_{Be})$; $N_\parallel = k_\parallel c/\omega$ is the component of the refractive index along the external magnetic field; $N_\perp^2 = c^2(k^2 - k_\parallel^2)/\omega^2$; ω_{pe} and ω_{pi} are the electron and ion plasma frequencies; and ω_{Be} and ω_{Bi} are the electron and ion cyclotron frequencies.

Equation (4.1) describes a slow wave which is practically longitudinal (E ∥ k) when $N_\parallel^2 \gg 1$. In this limit the wave is often referred to as an oblique Langmuir wave. It can propagate only in the plasma region where $\eta < 0$ and $\epsilon > 0$. When $\epsilon \neq 0$, $N_\perp^2$ is on the order of $N_\perp^2 \sim \omega_{pi}^2 N_\parallel^2/\omega_{Bi}^2 = c^2 N_\parallel^2/v_A^2$ (where v_A is the Alfvén velocity), so that the wavelength of this wave perpendicular to the magnetic field is extremely small, a fact which makes it possible to neglect the actual geometry of the tokamak.

Since the poloidal magnetic field in a tokamak is small, $k_\parallel$ may be regarded as directed along the toroidal magnetic field. Then, given the toroidal homogeneity of the system and neglecting its toroidicity, we can set $N_\parallel$ = const, so that the transverse refractive index of the slow wave goes to infinity when $\epsilon = 0$ (i.e., when the lower hybrid resonance condition is satisfied). Near the resonance the cold plasma approximation fails. It turns out [37] that when the thermal motion is taken into account, this slow mode undergoes linear conversion into a plasma wave which propagates in the direction of decreasing plasma density. For a finite plasma temperature, the conversion region is shifted slightly in the direction of lower densities relative to the location of the hybrid resonance [94]:

$$\epsilon^2 \simeq 6\beta_i^2 N_\parallel^2 \frac{\omega_{pe}^2\,\omega_{pi}^2}{\omega^4}\left(1 + \frac{1}{4}\frac{T_e}{T_i}\frac{\omega^4}{\omega_{Be}^2\,\omega_{Bi}^2}\right), \qquad (4.3)$$

where T_e and T_i are the electron and ion temperatures, $\beta_i = v_{Ti}/c$, and $v_{Ti} = \sqrt{2T_i/m_i}$ is the ion thermal velocity. The plasma wave that results from this conversion is slowed down strongly. In terms of the linear approximation, the damping of the wave is related to resonances at harmonics of the ion cyclotron frequency $\omega = s\omega_{Bi}$ (where s is an integer) which are always encountered in tokamak plasmas because of the magnetic field inhomogeneity.

The fast wave described by Eq. (4.2) can propagate in that part of the plasma where

$$\omega_{pe}^2 / (\omega | \omega_{Be} |) > N_\parallel^2 - 1.$$

Equations (4.1) and (4.2) are valid over the entire plasma layer only when the condition [42]

$$N_\parallel^2 > 1 + (\omega_{pe}^2 / \omega_{Be}^2)_{LH} \tag{4.4}$$

is satisfied, where $(\omega_{pe}^2/\omega_{Be}^2)_{LH} = \omega_{pe}^2/\omega_{Be}^2|_{\epsilon=0}$. When the opposite inequality is satisfied, points where the fast and slow waves undergo mutual conversion appear in the plasma layer (see Fig. 2.14) and the hybrid resonance zone is separated from the plasma edge by a fairly wide opacity region under actual conditions.

Equation (4.4), therefore, is of special significance in the LHH problem, as it determines the limits of accessibility to the resonance layer for waves launched from the plasma edge. It is easy to understand that there is also an upper limit to the accessible values of $N_\parallel$. For excessively high $N_\parallel$, Landau damping may be so strong that the rf energy will be absorbed at the plasma edge and the heating efficiency lowered [95]. A rough estimate of the absorption of the slow wave can be obtained with Eq. (4.1) by including the appropriate imaginary term η'' in η for Landau damping:

$$\eta'' = i \cdot 2 \sqrt{\pi} \; \frac{\omega_{pe}^2}{\omega^2 \, N_\parallel^3 \, \beta_e^3} \, \exp \left[- \frac{1}{N_\parallel^2 \, \beta_e^2} \right] ,$$

where $\beta_e = v_{Te}/c$ and v_{Te} is the electron thermal velocity. Finding the imaginary part of the refractive index $N_\perp''$ from Eq. (4.1) and setting $N_\perp'(\omega a/c) < 1$ (where a is the minor radius of the tokamak), to within logarithmic accuracy we obtain

$$N_\parallel^2 < 25/T_e , \tag{4.5}$$

where T_e is in keV.

In general, wave propagation and absorption under actual conditions in a tokamak are considerably more complicated. Nevertheless, the criteria (4.4) and (4.5) provide a qualitative description of the requirements on the longitudinal slowing-down spectrum of the waves excited by the antenna system. Several such systems for exciting waves of the necessary polarization with the required $N_\parallel$ spectrum have been proposed and experimentally tested. The most extensively studied systems are the so-called "grill" systems which consist of a set of parallel rectangular wave-guides with open ends directed toward the plasma column. A linear electrodynamic theory for these systems has been developed [96, 97]. In this theory the field in each waveguide is represented as a superposition of the waveguide mode with specified phase and amplitude incident on the plasma and the set of all reflected modes with unknown coefficients. The field at the plasma edge is the superposition of the solutions of the wave equation containing only waves which penetrate the plasma. Matching these fields at the waveguide cutoff yields a system of equations for the amplitude of the reflected waves. By solving these equations on a computer it is possible to obtain the $N_\parallel$ spectrum of the excited waves and a corresponding expression for the total reflectivity R_f for the incident rf energy.

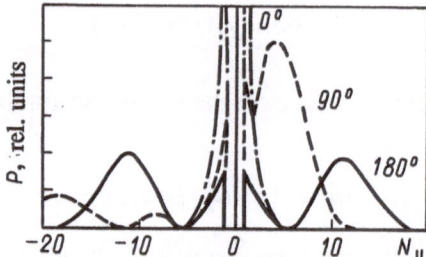

Fig. 4.1. The $N_{\parallel}$ spectral composition of the emission from a four-waveguide grill with different phasing for the waveguides [99].

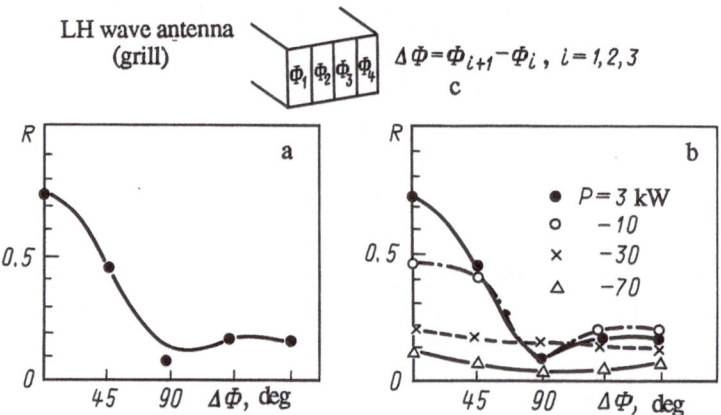

Fig. 4.2. The dependence of the power reflection coefficient on the phase shift between waves in neighboring waveguides [100]: a) calculation (solid curve) and experimental data (points) from the JFT-2 tokamak for $P = 3$ kW; b) experiments on JFT-2 at different rf power levels; c) grill arrangement.

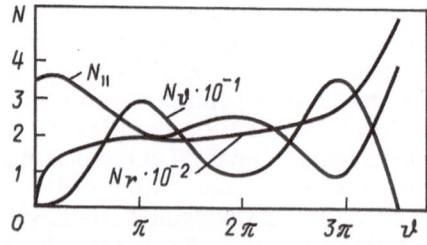

Fig. 4.3. The dependence of the parallel ($N_{\parallel}$), azimuthal (N_{ϑ}), and radial (N_r) components of the refractive index on the poloidal angle ϑ along a ray trajectory [105].

For a given antenna system geometry, the $N_\parallel$ spectrum depends primarily on the phase shift $\Delta\Phi$ between the incident waves in neighboring waveguides. The reflectivity also depends on this phase shift and has a minimum when the condition [98]

$$N_{\parallel g} \simeq \frac{15}{f(b-d)} \frac{\Delta\Phi}{\pi} > 1 \qquad (4.6)$$

is satisfied, where $N_{\parallel g}$ is a characteristic value of the refractive index, f is the frequency (GHz), b is the width of the waveguide (cm), and d is the distance between waveguides (cm). For a fixed $\Delta\Phi$ that satisfies Eq. (4.6), the reflectivity is determined mainly by the density gradient dn_e/dx of the plasma in the region near the grill and reaches a very flat minimum when

$$dn_e/dx \simeq 2\cdot10^{11} \, (N_{\parallel g}^2 - 1)^{1/2} f^3.$$

As an illustration, Figs. 4.1 and 4.2 show the results of computations and experimental data from a four-waveguide system on JFT-2 [99, 100].

Besides depending on the reflectivity, the operating efficiency of a grill depends strongly on the fraction of the power emitted outside the $N_\parallel$ range determined by Eqs. (4.4) and (4.5). This fraction is related to the number of waveguides and to the relationship among the parameters b, d, and c/ω. In order to optimize the system from this standpoint, it has been suggested that the grill be supplemented with passive elements (e.g., by corrugating the vessel walls [101] or using passive waveguide segments [102]).

Waves that satisfy Eq. (4.4) have one extremely interesting property. As can be seen from Eq. (4.1), when $N_\parallel^2 \gg 1$ the angle between the group velocity of these waves and the magnetic field in a dense plasma is very small and is practically independent of the refractive index. Thus, the waves excited by the antenna form two narrow beams (with a dimension along the magnetic field on the order of the grill size) oriented almost parallel (or antiparallel) to the external magnetic field. In other words, in the plane layer model the waves excited by the antenna form a structure similar to a resonance cone [103].

If we attempt to apply this conclusion to a tokamak geometry, the following picture unfolds: the waves, following the total magnetic field, go around the major and minor circumferences of the torus and penetrate the center of the plasma column relatively slowly. Clearly, the slow "burial" of the waves emphasizes the weak toroidal inhomogeneity of the system, which leads to significant changes in $N_\parallel$ along the ray trajectory of the wave. This fact may greatly alter the conditions for accessibility of the resonance, as well as the entire picture of wave propagation compared to the plane model.

Because of the complicated inhomogeneity of the plasma and magnetic field, wave propagation in tokamaks has been analyzed only by means of numerical calculations of ray trajectories [104–108]. These calculations have demonstrated the importance of toroidal effects (Fig. 4.3). Even when the initial values of $N_\parallel$ satisfy the condition (4.4), it is clear that at one of the minima of $N_\parallel$ along the trajectory an initial slow wave may be converted into a fast mode which propagates toward the edge. The latter wave, after being reflected from a low-density region, may be converted again into a slow wave as it propagates toward the center, and so on. Under certain conditions a slow wave may be reflected without conversion into a fast mode.

The resulting picture is fairly complicated and extremely sensitive to the plasma density and temperature profiles, as well as to the aspect ratio and q. This picture becomes significantly simpler only for initial values of $N_\parallel$ sufficiently large that linear conversion of the incident wave into a plasma mode occurs after the trajectory has "crossed" the minor radius once (without intermediate conversion into a fast mode and without reflections). In this case, however, condition (4.5) is often violated and Landau damping takes place comparatively far out in the plasma periphery.

Of course, ray tracing calculations are still not enough to obtain the structure of the wave field in a tokamak. It can merely be noted that this field should fill the entire plasma cross section between the critical density surfaces $\eta = 0$ and the hybrid resonance surface $\epsilon = 0$. The fast waves which result from conversion can even penetrate into the region contained within the lower hybrid resonance surface [104]. Resonance cones can be formed only at the plasma edge near the antenna.

Scattering of the waves on low-frequency oscillations in the plasma edge region has an important effect on their propagation [109]. This scattering produces significant poloidal components at the wave vector even in the plasma edge. As they propagate into the plasma these components grow roughly in inverse proportion to the minor radius and cause large changes in $N_\parallel$.

The influence of nonlinear effects on the propagation of lower hybrid waves is an extremely complicated question [110–113]. Estimates show that various kinds of decay instabilities can develop in homogeneous plasmas at present and planned power levels. Under actual conditions the plasma inhomogeneity usually raises the thresholds for resonance instabilities to safe levels, so that it is usually assumed [110–112] that stimulated scattering of the original wave on electrons or ions plays a dominant role. The threshold for such a process is given by [113]

$$P_{th} \simeq \left(\frac{Z_i}{A}\right)^{1/2} \frac{12,4}{N_\parallel} \left(1 + \frac{3\,T_i}{Z_i\,T_e}\right)^{1/2} \left(\frac{B_0}{5,0}\right)\left(\frac{T_e}{100}\right)^{3/2}\left(\frac{10^{12}}{n_e}\right) , \qquad (4.7)$$

where P_{th} is the threshold power (MW), Z_i and A are the charge and mass number of the ions, T_e is in eV, and n_e is in cm^{-3}.

In deriving Eq. (4.7) important use was made of the assumption that the pump wave field is localized in the resonance cone region. In this calculation convective removal of the energy of the scattered waves is also the mechanism that determines the instability threshold. For this reason Eq. (4.7) is obviously only applicable to the plasma edge, where toroidal effects have not caused the cones to "smear out."

It follows from these calculations [111, 112] that when the threshold power is raised by several times, wave transformation on particles leads to absorption of the applied power at the edge over a distance considerably smaller than the minor radius of the plasma. Saturation of the instability in this case is caused by cascade processes and occurs at a level when the energy in the scattered waves is comparable to the energy of the pump wave. As the rf energy dissipates, most of it apparently goes into suprathermal electrons [112].

Although a detailed comparison of the preceding theory with experiment is difficult because of the absence of reliable data on the plasma parameters in the edge region, crude estimates show that the rf power levels in most experiments are comparable to the threshold given by Eq. (4.7). Thus, it seems that, depending on the specific experimental conditions, regimes with strong parametric absorption of the rf energy in the plasma edge or regimes that are favorable to heating without strong instabilities may exist. In the latter case, quasilinear effects may have a strong influence on energy absorption.

At sufficiently high rf power levels quasilinear effects may significantly modify the picture of Landau damping on the electrons [114, 115]. As usual, a resonant interaction of the electrons with the waves leads to the formation of a plateau in the distribution function, so that noticeably reduced rf absorption appears to be caused by collisions which tend to Maxwellianize the electron velocity distribution function. According to a simplified one-dimensional theory [114], the power absorbed per unit plasma volume in the quasilinear regime is given by

$$p_e = \frac{2 + Z_i}{2\sqrt{2\pi}} \, \nu_{ei} n_e T_e \exp\left(-\frac{c^2}{v_{Te}^2 \, N_{\parallel 1}^2}\right) \ln \frac{N_{\parallel 1}}{N_{\parallel 2}}, \tag{4.8}$$

where $N_{\parallel 1}$ and $N_{\parallel 2}$ are the upper and lower bounds of the wave spectrum in $N_\parallel$ and ν_{ei} is the frequency of collisions between the background electrons and the ions.

It is, unfortunately, difficult to use this simple formula for directly interpreting experimental data because of the plasma inhomogeneity and, especially, because of changes in the spectrum owing to toroidal effects. An analysis of quasilinear absorption in a simplified plane layer model (neglecting changes in $N_\parallel$) [116] yields a weakened form of the criterion (4.5) which ensures the absence of absorption at the edge:

$$N_\parallel < 6.4/\sqrt{T_e}. \tag{4.9}$$

The results described here refer to the case when there are no waves with $N_\parallel > N_{\parallel 1}$ in the plasma. In the case of a rather slowly falling (as $N_\parallel \to \infty$) spectrum, on the other hand, quasilinear effects can enhance Landau damping and lead to absorption of rf energy at the edge with formation of a group of very hot electrons [117].

Naturally, with an asymmetric $N_\parallel$ spectrum, quasilinear distortions in the electron distribution function (formation of a plateau) will lead to the appearance of a directed electron velocity, i.e., a current in the plasma [114–116, 118]. The physical mechanism for this phenomenon is perfectly clear and involves the transfer of momentum from the wave to the electrons. Without examining it in detail, we note that according to some calculations [115, 116] the efficiency of current drive by means of lower hybrid waves is fairly high even under reactor conditions.

The energy of lower hybrid waves is evidently absorbed by ions through so-called stochastic heating under real conditions [58]. (This mechanism was discussed in detail in Section 2.13.) It is closely related to the phase resonance of an ion in the wave field:

$$\omega = k\mathbf{v}. \tag{4.10}$$

When the ion moves in its cyclotron orbit, the resonance (4.10) can be satisfied only at certain times, near which the ion gains (or loses) some energy from (to) the wave. For a small amplitude wave with $\omega \neq s\omega_{Bi}$ (with s an integer) the time intervals over which the ion is accelerated by a wave alternate regularly with intervals over which it is slowed down, so that on the average there will be no energy exchange between the ions and a wave. When the condition

$$E/B_0 > (1/4)\,(\omega_{Bi}/\omega)^{1/3}\,(1/N_\perp) \tag{4.11}$$

is satisfied, however, the perturbation of an ion's cyclotron orbit by the wave causes the phases of the successive arrivals of the ion in the resonance region (4.10) to vary strongly. As a result, a regime develops in which the ions are almost randomly accelerated or slowed down by the wave at the resonance (4.10). The ions will then diffuse in velocity space in accordance with Eq. (2.226) and undergo stochastic heating.

From this discussion it is clear that only those ions whose perpendicular velocity is greater than the phase velocity of the wave will undergo stochastic heating. Thus, the waves must be strongly slowed down for efficient absorption of the rf energy [116]:

$$\omega/(k_\perp \mathbf{v}_{Ti}) < 2\sqrt{2}. \tag{4.12}$$

For typical experimental conditions this criterion is satisfied near the region where the slow lower hybrid wave is converted into a plasma mode. Since slowing down of the lower hybrid wave becomes greater in the direction from the plasma edge to the conversion region, the suprathermal ions in the tail of a Maxwellian distribution begin to heat up first. When condition (4.11) is satisfied, a plateau develops in the perpendicular velocity distribution of the ions owing to stochastic ion heating. Collisions cause tilting of this plateau, so that stochastic heating results in a quasi-Maxwellian "tail" of energetic ions with an effective temperature [58]

$$T_{ih} = T_i \left[1 + \frac{2^{3/2}\,a_0\,E^2}{3\,n_i\,T_i}\,\frac{\omega_{pi}^2}{\omega\nu_{ii}}\left(\frac{\omega}{k_\perp \mathbf{v}_{Ti}}\right)^3 \right], \tag{4.13}$$

where ν_{ii} is the ion–ion collision frequency and a_0 is the ratio of the area of the magnetic surface intersected by the rf waves to the area of the entire magnetic surface.

In this model bulk ion heating is caused by collisions between hot ions and background ions. The power transferred to unit volume of the plasma is then given by

$$
P_i = \frac{3}{2\sqrt{2}}\, n_i m_i \nu_{ii} \nu_{Ti} \left(\frac{c}{N_{\perp_2}} - \frac{c}{N_{\perp_1}} \right) \exp\left[- \frac{c^2}{N_{\perp_1}^2 \, \nu_{Ti}^2} \right], \tag{4.14}
$$

where $N_{\perp 1}$ and $N_{\perp 2}$, the maximum and minimum values of the perpendicular refractive index of the waves, respectively, determine the extent of the quasilinear plateau in the ion distribution function: $\nu_1 = c/N_{\perp 1}$ and $\nu_2 = c/N_{\perp 2}$.

Equations (4.14) and (4.13) have been derived neglecting the actual inhomogeneity of the plasma and magnetic field in tokamaks. This inhomogeneity can affect the results both through the wave characteristics [the $N_\perp$ spectrum, $E(N_\perp, \mathbf{r})$, etc.] and because of ion drift along banana orbits. Because of this drift the particles are in the stochastic heating region (near the conversion region) for only a limited time and the hot ions are efficiently mixed over the cross section of the plasma column. Numerical modelling [119] of this process has shown that fast ion heating is also stochastic and occurs only after many transits of the ions along banana orbits. Unfortunately, these results [119] were not adequate for empirical generalizations and establishing quantitative criteria for evaluating the role of this factor in ion heating under different conditions.

The above theoretical discussion of the processes that occur during LHH of tokamak plasmas can be briefly summarized as follows: efficient plasma heating is possible only when the thresholds for parametric instabilities are not greatly exceeded at the power levels being used. Then ion heating is apparently by a stochastic mechanism and takes place when the linear conversion region is localized in the central region of the plasma. Quasilinear Landau damping is responsible for electron heating. Depending on the $N_\parallel$ spectrum of the waves, heating may occur at lower densities, as well as at the maximum plasma density satisfying the conditions for linear conversion. The energy deposition profile during rf heating depends strongly on the $N_\parallel$ spectrum of the waves, which can in turn be affected strongly by toroidal effects and scattering on density fluctuations.

At higher power levels the heating efficiency appears to be low because of parametric instabilities that lead to absorption at the plasma edge.

A range of factors, therefore, can affect LHH. It is rather difficult to take them all into account self-consistently. Hence, until recently only very simplified models have been used for following the contribution of a given effect in isolation. For example, considerable attention has been devoted to numerical computations of the quasilinear evolution of the ion and electron distribution functions during heating in a uniform plasma model [55, 58, 115] or with the plasma inhomogeneity taken into account in a simplified manner [106, 116]. Several codes have been developed for numerical computation of the ray trajectories of lower hybrid waves in tokamaks [104–108, 121, 122] and of the characteristics of antenna systems [96, 97, 123–124]. Direct numerical experiments have been done [125–127] in which the plasma was replaced by a fairly large number of "particles" (on the order of 10^4 [125]) located in a uniform magnetic field. The particles were treated in a somewhat simplified fashion with an unrealistic ratio of the electron and ion masses and the particles were regarded as "two dimensional" [125] (i.e., all variables depended only on two spatial coordinates). It was assumed that the particles interact according to Coulomb's law and the exact equations of motion were solved. The initial distribution of the particles represented a plane plasma layer with a particle density that increased from the boundary. An antenna was simulated by external charges located on the plasma boundary.

In these numerical experiments electron heating and the formation of a group of fast ions in the interior and at the edge of the plasma were observed, as well as significant modifications of the density profile near the antenna. The ion heating is evidently associated with a stochastic mechanism. Under these experimental conditions the wave was usually absorbed in a region where it had not been slowed down enough to heat "thermal" ions (ions from the bulk of the distribution function). Electron heating was quasilinear in nature, but effects owing to particle trapping by the wave showed up clearly. It is interesting to note that linear wave conversion was not observed in the numerical simulations, apparently because the waves were strongly damped before reaching the transformation region. The approximation of weak spatial dispersion was violated in the conversion region for the parameters used in the calculations, i.e., $k^2 v_{Ti}^2/\omega^2$ was not small. Under these conditions in a model with unmagnetized ions the plasma wave is strongly damped. It can also be efficiently absorbed stochastically.

Although numerical simulations are extremely interesting, because of simplifications in the calculations they serve more to illustrate the main physical phenomena during heating than to model the actual situation. Attempts have been made to take more realistic account of the properties of plasmas in tokamaks [128–130]. These models include calculations of the time evolution of the plasma parameters [$T_e(r, t)$, $T_i(r, t)$, $j(r, t)$] with the aid of a radial transport code. The results are used to compute the equilibrium magnetic field configuration. Then the ray trajectories of the waves are calculated in a real toroidal geometry, with the parallel wave number ($N_{\parallel}$) spectrum of the antenna system broken up into a number of segments (74 in [128]), each of which is associated with a separate ray trajectory.

The interaction of the waves with the electrons at each magnetic surface is described by a local (neglecting spatial derivatives) Fokker–Planck equation in which the effects of the induced electric field and the wave field on the electron distribution function are taken into account by means of the $N_{\parallel}$ distribution that is calculated with the aid of the ray trajectories. Losses of fast electrons from a given magnetic surface are also taken into account phenomenologically. The absorption of waves by ions has been described in terms of the linear theory (in a model of unmagnetized ions) [128] and in terms of a quasilinear theory of stochastic heating [130]. The whole system of equations is self-consistent in the sense that the quasilinear evolution of the distribution function on a given magnetic surface is calculated using the $N_{\parallel}$ spectrum obtained by taking into account the damping of the waves along their ray trajectories which, in turn, are calculated with the evolution of the distribution function taken into account. Furthermore, the ray trajectories are calculated from profiles obtained using a transport model for which the sources are the energy deposition profiles calculated with the aid of the same ray trajectories, etc.

These models, of course, have certain fundamental simplifications. Thus, the possibility of parametric and modulational instabilities, as well as of wave scattering on plasma fluctuations, are not fully taken into account. The role of refraction and interference of waves in calculations of energy deposition profiles using ray trajectories is not clear [121]. The possible effect of rf waves on transport processes in plasmas is not modelled in any fashion. Instabilities that may develop when the electron distribution function is strongly distorted, and so on, are not taken into account. Nevertheless, a quantitative comparison of these model calculations with experiment is of great interest. It would undoubtedly clarify the relative contributions of the various processes that take place during LHH.

Up to now, however, the model studies have been primarily devoted to regimes in which currents are generated by the application of rf power to plasmas.

4.2. HEATING TECHNIQUES

The lower hybrid resonance frequencies characteristic of the parameters of tokamaks and other toroidal devices lie in the decimeter wavelength range, from 70 cm (for $B_0 \approx 1$ T, $n_e \sim 10^{13}$ cm^{-3}) to 7 cm (for $B_0 \approx 10$ T, $n_e \sim 10^{15}$ cm^{-3}). High-power microwave sources (klystrons and magnetrons) have been developed for these bands.

The power of modern klystrons is as high as 1 MW and their efficiency is 40–60% for pulse lengths on the order of 1 s even in a cw regime [131]. It can clearly be increased still further if necessary. The advantage of klystron amplifiers is the ability to control the phase, frequency, and power with a master oscillator (which operates at a low power level). At the same time, magnetron generators are simpler and smaller and require a lower supply voltage. They can also be used in an amplifier regime to control their parameters, although this involves some complications.

The techniques of decimeter wave transmission along coaxial and waveguide channels, as well as for making measurements in these channels, are well developed [132]. Here we shall dwell only on the methods of introducing microwave energy into a vacuum vessel.

The simplest system for exciting lower hybrid waves, which was used in the first tokamak experiments, is a structure made up of several loops [133, 134] that excite the magnetic field of the waves. The shape and location of the loops determines the longitudinal wave number spectrum. The deficiencies of these systems are a broad spectrum and a comparatively low maximum power (limited by electrical breakdown in the windings). In order to improve the spectrum in experiments on the FT-1 tokamak [134] a slowing-down structure was used. This structure consisted of limiters with circular apertures mounted at equal distances from one another inside the bellows vacuum vessel of the tokamak over 1/4 of its length. It substantially narrowed the wave number spectrum excited by the magnetic loops.

Grill systems are most widely used in LHH experiments [96, 135] (see Fig. 4.2). Their longitudinal slowing-down spectrum is determined by the number of waveguides and by the phase shift between the waveguides determined by the input system. The method for calculating the spectrum has been described elsewhere [96–98]. These calculations show that grill systems easily produce an optimum spectrum and low reflection coefficients. At low rf powers the experimental results agree well with calculations based on a reasonable model for the plasma density distribution near the waveguide boundaries (Fig. 4.2a). As the power delivered to the plasma is raised, however, this agreement is lost. Usually the reflection coefficient decreases and ceases to depend on the phase shift $\Delta\Phi$ (Fig. 4.2b). Two possible reasons for this effect have been proposed. One is related to the effect of the waves on the plasma parameters near the waveguides. Measurements on the Wega tokamak, for example, show a "flattening" of the density profile near the grill

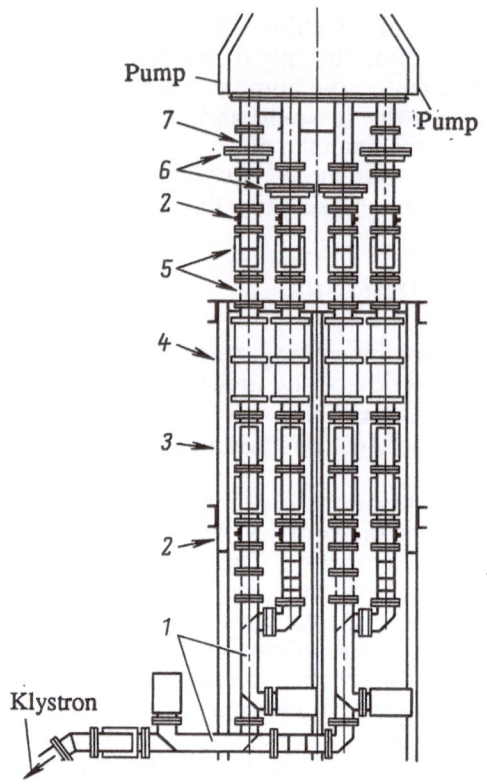

Fig. 4.4. The lower hybrid heating system on the Petula B tokamak [141]: 1) power divider; 2) coupler; 3) U-shaped waveguide; 4) ferrite valve; 5) flexible waveguide; 6) dc break; 7) vacuum window.

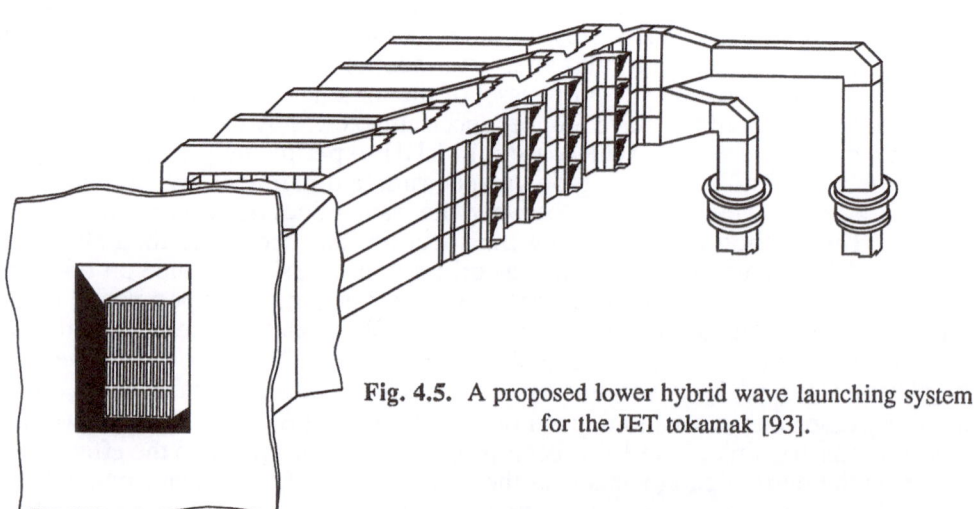

Fig. 4.5. A proposed lower hybrid wave launching system for the JET tokamak [93].

caused by the wave field [136]. A strong effect of rf power on the density and temperature profiles has also been demonstrated in a special simulation experiment on a linear machine [137]. The other, and possibly more important reason, is the appearance of electrons inside the waveguides. These electrons may be produced by a secondary emission (multipactor) discharge caused by the wave field and can cause substantial changes in the phase relationships at the waveguide outlets [138, 139].

In order to prevent secondary electron discharges, the waveguides in some experiments have been coated with materials that have a low secondary emission coefficient (titanium, carbon). In addition, special steps have been taken to outgas the surface of the power input system (gas discharge cleaning, gettering). These steps have made it possible to deliver rf power at densities of up to 5 kW/cm^2 [139]. Record power densities of 9 kW/cm^2 have been achieved in experiments on the Alcator C tokamak [140]. It should be noted that processes occurring at the boundary between the antenna system and the plasma may change the spectrum of the waves that are excited. So far, no systematic data on the spectrum in the presence of a plasma are available.

Figure 4.4 is a sketch of a typical grill system for exciting lower hybrid waves on the Petula B tokamak [141]. A grill system for the JET tokamak [93] that has been designed to deliver 5 MW at frequencies of 1.3–1.5 GHz in pulses lasting up to 10 s is sketched in Fig. 4.5. A still more subdivided system has been proposed for the INTOR tokamak [142]. Consisting of 240 waveguides, it is intended to deliver a power of 25 MW. Analyses show that the greatest difficulty in constructing antenna systems for large machines is associated with cooling them effectively.

4.3. EXPERIMENTAL STUDIES

The first experiments on lower hybrid plasma heating in tokamaks were done on the FT-1, TM-3, and ATC machines [143–146]. Lower hybrid studies have since been carried out on a whole series of tokamaks (Table 4.1). In most of these experiments, LHH is done at power levels on the order of the ohmic heating power. Only in the most recent experiments on the Alcator C and Petula B tokamaks has this level been significantly exceeded. We now discuss the most important results of these experiments.

As noted above, coupling efficiencies of 60 to 90% can be obtained with systems for exciting lower hybrid waves. With a grill waveguide system it is possible to form a longitudinal wave number spectrum at the plasma edge which is close to the optimum for LHH. Processes at the boundary between the exciting system and the plasma, however, can affect the spectral composition of the excited waves. Determining the spectrum and studying the propagation of the excited waves in a real tokamak plasma is one of the central problems in research on LHH. Until recently data on wave excitation in tokamaks were obtained with the aid of probes located in the limiter shadow. Direct data on wave penetration into the central hot region of a tokamak and on the wave characteristics can be obtained from millimeter or infrared scattering experiments. This kind of experiment has been set up to study LHH on Alcator A [146], Wega [148], and several other tokamaks. These experiments showed that the wave number spectrum of the excited waves does not differ greatly from the linear theory. In the experiments, however, there was no dependence of the spectrum on the phase shift between the waveguides in the

Table 4.1. Lower Hybrid Heating Experiments in Tokamaks

Machine	Refs.	Plasma parameters				Parameters of rf system			Launch
		a (cm)	B (T)	$\bar{n}$, 10^{-3} cm^{-3}	f (MHz)	P (kW)	τ (ms)		
FT-1	[143, 154]	15	1	1	400	80	3		"Comb"
Wega	[148, 151]	15	2.3	1–4	800	400	50		Loop, 4-grill
JFT-2	[138]	25	1–2	1–2	650	400	15		4-grill
Petula B	[152]	16	2.7	2–6	1250	800	100		2 × 4 grill
JIPP-TII	[156]	17	2.2	1–2	800	300	20		2 C-shaped waveguides
Alcator A	[150]	10	5.4	10–20	2500	90	20		2-grill
Alcator C	[129, 140]	16.5	8–10	5–20	4600	1100	60		4 × 4 grill
Versator II	[129]	13	1.4	3–5	800	100	5		4 × 6 grill
FT	[155]	20	6–8	5–20	2450	300	400		2 × 2 grill
PLT	[149]	40	2.7	2	800	400	150		6-grill
Asdex	[157]	40	2.5	1–5	800	1300	500		8-grill

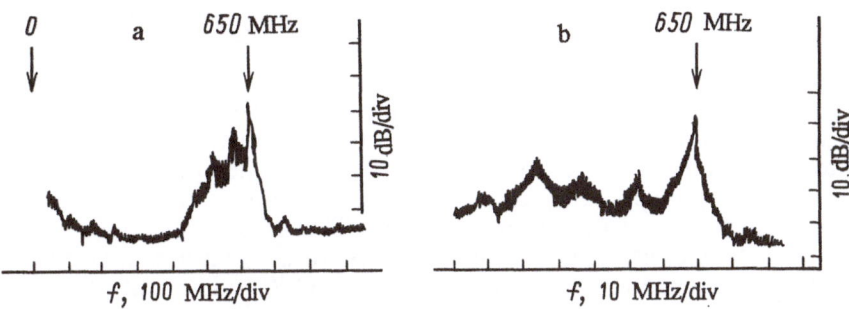

Fig. 4.6. The decay spectra of lower hybrid waves in the JFT-2 tokamak [138]. (The spectra were measured in the limiter shadow using probes.)

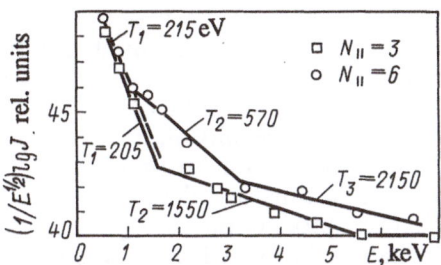

Fig. 4.7. The energy spectrum of charge exchange neutrals during LHH in the Wega tokamak [148] ($\bar{n} = 2.8 \cdot 10^{13}$ cm^{-3}, $P = 130$ kW).

launching system. Experiments indicate that waves penetrate into the central plasma region without any significant azimuthal nonuniformity in the wave amplitude. (Such a nonuniformity might be a consequence of resonance cones.) Some frequency broadening of the spectrum ($\Delta f < 5$ MHz) has been observed and attributed to scattering of the waves on plasma drift oscillations.

The excitation of parametric instabilities in the plasma edge has been reported in many papers. They were first observed during LHH on the ATC tokamak [145]. In a number of cases a reduction in heating efficiency has been noticed in association with them. The most detailed studies of edge instabilities and their effect on plasma heating have been done on the JFT-2 and FT tokamaks [100, 138, 147, 155]. The characteristics of the oscillations were measured in these experiments by probes located in the limiter shadow. Figure 4.6 shows the frequency spectra corresponding to two decay instabilities observed and identified on JFT-2. The first is related to the excitation of whistler waves and ion quasimodes (i.e., to stimulated scattering of the pump wave on ions). The second corresponds to resonance decay of the wave excited in the plasma into oblique Langmuir and ion cyclotron waves. The threshold for the decay instability depends strongly on the plasma density and temperature [138]. Measurements on FT [155] showed that the threshold power is determined by the ratio of the electron density to the electron temperature at the edge and is in satisfactory agreement with Eq. (4.9), according to which $P_{th} \sim (T_{eg}/n_g)^{3/2}$. This implies that the decay instability can be suppressed by increasing the edge temperature T_{eg}. This was shown in experiments on JFT-2 where edge heating accompanied heating of the main plasma. Suppression of the decay instability associated with edge heating was also observed when LHH was supplemented by neutral beam injection [138] and with ECH [82] in JFT-2.

The experiments on JFT-2 and FT have demonstrated a definite correlation between the ion heating efficiency in the center of the plasma and parametric instabilities: when the amplitude of the instabilities fell, the heating efficiency rose. It is natural to attribute this to edge absorption of rf energy when parametric instabilities develop and to a corresponding reduction in the penetration of lower hybrid waves into the central region of the plasma column. This effect was confirmed experimentally in LHH experiments on Alcator C where CO_2 laser scattering on lower hybrid waves was used to locate the fluctuations [129]. The scattering measurements showed that when a strong parametric instability is present the lower hybrid wave does not penetrate into the plasma center but is absorbed at the edge where the decay processes are localized. In a number of experiments it has been noted that groups of fast ions, whose origin is linked to decay processes, are formed at the plasma edge simultaneously with edge absorption. In the JFT-2, for example, a direct correlation was established between the number of fast ions and the width of the detected decay spectrum. The confinement time for these ions is usually short and close to their drift time along toroidal orbits.

We now examine experimental results on ion heating in the main plasma volume. In almost all experiments where ion heating has been observed, the formation of hot "suprathermal" groups of ions was detected when rf power was applied to the tokamak plasma. Fast ion production during LHH was fist discovered in experiments on the FT-1 tokamak [143], where it was also found that a substantial fraction of these ions have a confinement time much greater than the drift time and on the order of the confinement time for the bulk ions.

Experimental data on fast ion formation are obtained by measuring the energy spectrum of charge exchange atoms emitted from the plasma and the neutron yield. The energy spectra of charge exchange atoms can usually be represented as the superposition of two Maxwellian distributions (e.g., Fig. 4.7). Scans made with a charge exchange neutral energy analyzer in several experiments have shown that a fast ion source with a confinement time greater than the drift time is located in the plasma core. This is also indicated by measurements of the neutron flux produced by the fast deuterium ions [149, 150].

A detailed investigation of the conditions for fast ion production during LHH has been carried out on the Wega tokamak [148, 151]. Two types of grill systems were used which could excite waves with different longitudinal slowing-down spectra. The charge exchange neutral spectra and, therefore, the ion energy distribution functions were different for the two systems (Fig. 4.7). In agreement with the theory of stochastic quasilinear heating, the lower bound on the energy of the fast ions is fairly consistent with the maximum values of the perpendicular refractive indices in the linear conversion region obtained with the plane layer model from computed $N_{\parallel}$ spectra. The plasma density dependence of fast ion production efficiency is also in agreement with the theoretical model. It was found that charge exchange neutrals with a given energy appear only when the density in the plasma column reaches a level such that waves with the necessary transverse slowing down should exist. This fact is illustrated in Fig. 4.8, which shows the dependence of the transverse refractive index $N_{\perp}$ on the plasma density, calculated using a plane layer model, and the ion energy $E_{\perp} = m_i c^2/(2N_{\perp}^2)$ corresponding to this refractive index. The shaded rectangles and diamonds represent the lower (with respect to the density) limits at which charge exchange neutrals with the given energy have been

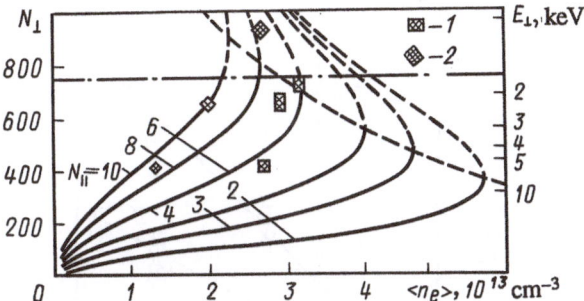

Fig. 4.8. The perpendicular refractive index and limiting energy of fast charge exchange neutrals in the Wega tokamak [148]: the limiting energy E_b (shaded rectangles) is shown for grills with different longitudinal refractive indices (slowing-down): 1) $N_{||} = 3$; 2) $N_{||} = 6$.

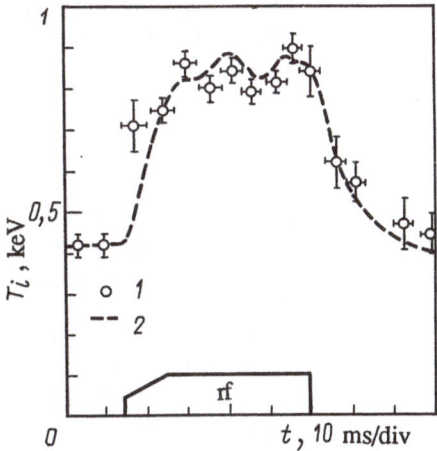

Fig. 4.9. The change in the ion temperature during lower hybrid heating in Petula B [152]: 1) charge exchange; 2) neutron flux measurements.

observed. They appear to be localized near the dispersion curves $N_\perp^2$, corresponding to different amounts of longitudinal slowing down, $N_{||} = 10$ and $N_{||} = 6$. These values of $N_{||}$ are close to the limits of the computed spectra for the grills that were used. This agreement, however, is not observed in all experiments. A strong discrepancy was observed, for example, in the FT tokamak, where the longitudinal slowing down corresponding to the limiting ion energy was higher than the limits for the spectrum excited at the boundary. The authors relate this discrepancy to an increase in the maximum $N_{||}$ when lower hybrid waves propagate in the plasma. It may be caused by the two-dimensional character of the plasma inhomogeneities or by scattering of the waves.

A comparison of the experimentally observed fast ion temperature with Eq. (4.13) is difficult because of the lack of reliable estimates for a_0, E, and $\omega/(k_\perp v_{Ti})$. If, however, in this formula we substitute the value of $k_\perp$ corresponding to the

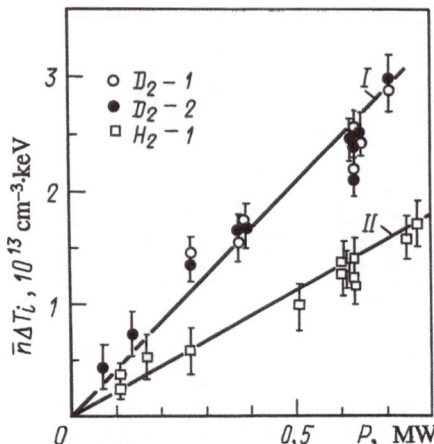

Fig. 4.10. The dependence of the ion temperature increase on the power during lower hybrid heating in Petula B [152]: I) $H_2, \bar{n} = 4.5 \cdot 10^{13}$ cm^{-3}; II) $D_2, \bar{n} = 6 \cdot 10^{13}$ cm^{-3}; 1) charge exchange; 2) neutron flux measurements.

conversion region, take the value of E near the grill, and estimate a_0 as the ratio of the product of the grill length multiplied by 1/8 of the poloidal perimeter of the plasma column to the entire surface area, then the result will be more or less of the same order as the data from a whole series of experiments [151].

Heating of the bulk ions has been observed in many LHH experiments, beginning with experiments on the ATC and Wega tokamaks. The greatest amount of heating has been obtained on the Petula B tokamak [152]. A power of up to 800 kW was applied to the plasma and a heating efficiency of up to $\eta_h = 4 \cdot 10^{13}$ eV/(kW·cm^3) was obtained. Figures 4.9 and 4.10 show some results from the Petula B experiments on the heating dynamics including measurements of the charge exchange neutral spectrum, Doppler broadening of impurity ion lines, and neutrons. Figure 4.10 illustrates the linearity of the heating effect and shows how the efficiency increases with density. Heating was accompanied by an increase of 2–3 times in the impurity density (Z_{eff} rose from 2 to 3); however, the radiation losses were still considerably lower than the heating power.

Experiments on various tokamaks are in qualitative agreement with the idea that the bulk ions are heated by energy transfer from a group of ions that has been heated stochastically. The conditions for this type of heating depend on the ratio of the minimum perpendicular phase velocity of the wave to the ion thermal velocity,

$$\delta_i = v_{ph \perp min} / v_{Ti} \approx (2E_b / T_i)^{1/2}.$$

This dependence has been derived [153] from published experimental data (Fig. 4.11). It is clear that significant heating is obtained when $\delta_i < 3.5$–4, which is consistent with numerical calculations. Heating through a group of hot ions is also confirmed by the strong (quadratic) dependence of the heating efficiency on plasma density.

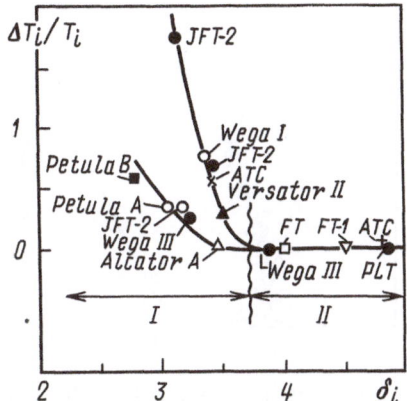

Fig. 4.11. The dependence of ion heating on the parameter δ_i in different LHH experiments [153]: I) ion heating region; II) wave–particle interaction region.

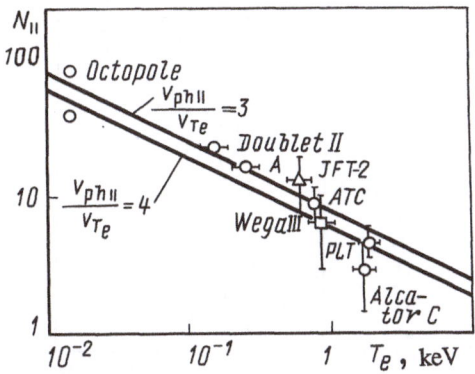

Fig. 4.12. The dependence of the longitudinal slowing-down (parallel refractive index) on the electron temperature in different LHH experiments where electron heating has been observed [153].

In addition to this model, several papers (e.g., [138]) have examined the possibility of direct heating of the bulk ions. The specific mechanism for this heating remains unclear. It should also be noted that bulk ion heating has by no means been obtained in all experiments. It did not occur, in particular, in experiments on the Alcator C and PLT tokamaks. In some machines, this type of heating occurred, but not always. The most probable reason for this is strong edge absorption caused by parametric instabilities. The critical dependence of heating on the experimental conditions can be explained by the influence of the edge plasma parameters in driving these instabilities. This explanation is in good agreement with the measurements of wave penetration into the central plasma region of Alcator C.

A strong dependence of the ion heating efficiency on the position of the antenna has been observed in experiments on FT-1 and it has been shown that this dependence is caused by toroidal drift of fast ions formed in the zone adjacent to the antenna [80]. Depending on the antenna's location relative to the drift direction, the

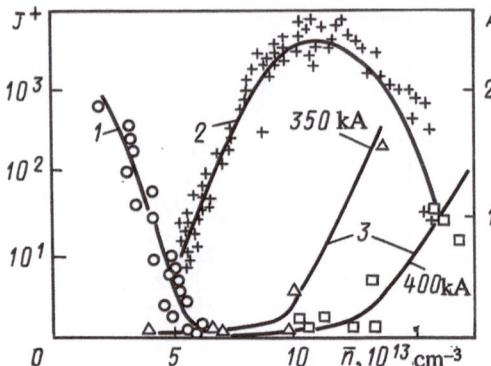

Fig. 4.13. The conditions for plasma heating in the FT tokamak [155]: 1) amplitude of emission at $\omega = \omega_{Be}$ (scale A); 2) fast neutral flux (scale J); 3) amplitude of the decay spectrum (scale A).

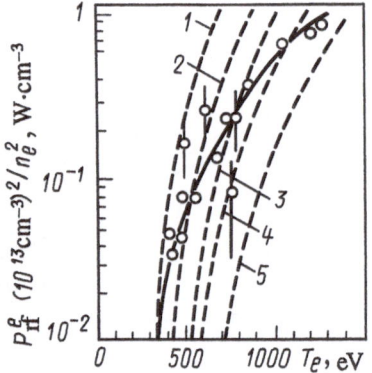

Fig. 4.14. The specific power absorbed by electrons during LHH in the Wega tokamak [151]. [The dashed curves are theoretical curves (4.8) obtained for different parallel slowing-down spectra.] 1) $N_{\|1} = 9$, $\alpha = 2$; 2) $N_{\|1} = 8$, $\alpha = 2$; 3) $N_{\|1} = 7$, $\alpha = 4$; 4) $N_{\|1} = 7$, $\alpha = 2$; 5) $N_{\|1} = 6$, $\alpha = 4$. ($\alpha = N_{\|1}/N_{\|2}$, $N_{\|1}$ and $N_{\|2}$ are the limits of the $N_{\|}$ spectrum.)

fast ions may be lost to the wall or drift inside the plasma, transferring their energy to the bulk ions. In the latter case the heating parameter for the ions is 3–4 times that in the former. In the FT-1 tokamak it reached $4 \cdot 10^{13}$ eV/(kW·cm^{-3}).

We now consider the experimental data on electron heating. The first data on the formation of groups of fast electrons by lower hybrid waves was obtained on the TF-3 tokamak [144]. Bulk electron heating was first observed on FT-1 [154]. Further experimental studies have shown that electron heating usually occurs at densities below those at which the ions are heated (i.e., at densities considerably lower than those corresponding to the lower hybrid resonance). Analysis showed that the experimental results are mostly consistent with electron heating by Landau damping including quasilinear relaxation. This type of heating is determined by the ratio of the parallel phase velocity of the wave, $v_{\text{ph}\parallel}$, to the electron thermal velocity v_{Te},

$$\delta_e = v_{\text{ph}\parallel}/v_{Te} = (c/N_\parallel)\sqrt{m_e/T_e}.$$

Figure 4.12 [153] shows data from experiments in which electron heating was distinctly observed. The vertical lines at the plot points represent the range of values of $N_\parallel$ corresponding to the width of the principal maximum of the excited spectrum. This range overlaps the interval $\delta_e = 3$–4 in accordance with the quasilinear heating theory. A significant deviation was noticed in a later experiment on the FT tokamak [155] in which $N_\parallel$ lay below the curve for $\delta_e = 3$–4. The authors explain this by an increase in $N_{\parallel\,\text{max}}$ as the waves propagate in the plasma. The density boundary between the electron and ion heating regions can be determined from the approximate equality $\delta_e \approx \delta_i$. At densities above this boundary the lower hybrid waves are absorbed by fast ions and do not reach the region where electron absorption is strong. With the aid of the dispersion relation this equality yields

$$n_g = n_c T_e \, \omega_{Be}^2 / [\omega_{Bi}|\omega_{Be}|(T_e + T_i) - \omega^2 T_e].$$

This formula agrees with experimental data on the threshold for ion heating in various tokamaks. A direct observation of this transition in experiments on the FT tokamak [155] is illustrated by Fig. 4.13 which shows the variations with plasma density of the intensity of cyclotron emission from fast electrons and of the flux of energetic charge exchange atoms (which characterizes the delivery of energy to the plasma ions). It is clear that at densities of $(3$–$7)\cdot10^{13}$ cm^{-3} there is a transition from electron heating to ion heating. At densities of $(1.2$–$1.5)\cdot10^{13}$ cm^{-3} parametric instabilities develop and the ion heating efficiency decreases.

A sharp dependence of the heating efficiency on the electron temperature and a quadratic dependence on the density, both of which are characteristic of heating through Landau damping (Figs. 4.14 and 4.15), are confirmed experimentally. Experimental data on the T_e dependence of the specific power delivered to the bulk electrons during LHH in the Wega tokamak [151] (Fig. 4.14) are in satisfactory agreement with Eq. (4.8) for $N_\parallel = 6$–9. These values of $N_\parallel$ are close to the spectrum generated by the antenna system in Wega.

The strongest electron heating has been obtained in an experiment in Alcator C with magnetic fields of 8–11T, currents of up to 500 kA, plasma densities of $(0.8$–$1.7)\cdot10^{14}$ cm^{-3} and temperatures of 1–2 keV [129]. Heating powers of up to 1.1 MW at a frequency of 4.6 GHz were applied. Radiation from impurity atoms, whose entry into the discharge depended strongly on the limiter material, played a strong role in the heating process. The best results were obtained with carbon limiters coated with SiC. The dependence of heating in deuterium on the electron density is shown in Fig. 4.15. A group of fast electrons was detected during heating. The fraction in this group was on the order of 10^{-3}, which is in rough agreement with the quasilinear relaxation model. The "tail" and bulk electron heating both cease at densities corresponding roughly to the threshold beyond which fast ion heating occurs. Electron heating also decreases at densities below 10^{14} cm^{-3} because of reduced efficiency of energy transfer from the "tail," which has been heated by Landau damping, to the bulk electrons. As can be seen in Fig. 4.15, the electron temperature increase was as much as $\Delta T_e = 1.2$ keV. At the same time, ion

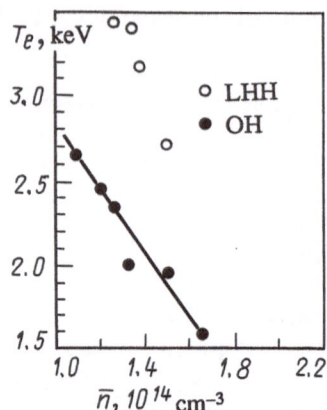

Fig. 4.15. The dependence of the electron temperature on the plasma density during LHH in Alcator C [129]: D_2, $B = 9$ T, $P = 850$ kW, Thomson scattering measurements.

heating with $\Delta T_i \approx 0.8$ keV, an electron density increase $\Delta n_e \approx 15\%$, and an impurity influx (Z_{eff} increased to 5) were observed. Simulations showed that the ion heating was not caused by direct transfer of energy from the rf field to the ions. It takes place through the electrons and is partially related to a reduction in the density of the main plasma ions because of the influx of impurities. As a consequence, a record high efficiency was achieved in this experiment: $\eta_h = 22 \cdot 10^{13}$ eV/ (kW·cm³). The numerical simulations showed that the fraction of the LHH power absorbed in the central region of the plasma ($r/a < 0.5$) was about 60%.

In addition to LHH, in recent years extensive work has been under way on the excitation of rf-driven currents in tokamaks by lower hybrid waves [55]. These currents develop as a result of the transfer of momentum to fast electrons during absorption of waves by Landau damping. They can be efficiently excited only at low densities ($n_e < 0.2$–$0.5 n_{LH}$) and the efficiency increases as the density is lowered ($I \propto n_e^{-1}$). Research on current drive is important for the tokamak program since it offers a way of creating a steady-state tokamak, but it is beyond the scope of this book.

4.4. SUMMARY OF RESEARCH

Despite the complexity of the processes responsible for heating in the lower hybrid range, theoretical and experimental research have led to an understanding of the overall heating behavior. It has been shown that three principal mechanisms are responsible for absorption of waves in the plasma and plasma heating: stochastic absorption, Landau damping, and absorption associated with decay instabilities. The stochastic interaction is effective for strong perpendicular slowing-down of the wave (i.e., large $N_{\parallel}$) and produces fast ions through which the bulk ions are heated. Landau damping produces fast electrons through which the bulk electrons are heated. Finally, under certain conditions decay instabilities are driven near the plasma boundary; these cause edge absorption and heating of ions which rapidly leave the plasma. Competition among these mechanisms usually determines the type of heating. Large perpendicular slowing-down occurs in the neighborhood of the lower hybrid resonance, i.e., at plasma densities close to the lower hybrid density (n_{LH}). At this density efficient ion heating has also been realized through stochastic interactions. At lower densities Landau damping is stronger and leads to electron heating. The transition between these two heating regimes usually occurs at $n_e \simeq 0.5 n_{LH}$. In some experiments edge heating and absorption owing to decay instabilities have been noticed. Sometimes they completely prevented penetration of the waves into the plasma center and heating of the core plasma. The threshold density for driving decay instabilities usually falls in the stochastic ion heating range, but it depends strongly on the parameters of the edge plasma.

The highest powers delivered to plasmas in the lower hybrid frequency range are 800 kW in Petula B and 1100 kW in Alcator C. An ion heating regime has been obtained in Petula B. At a density of $\bar{n}_e \sim 5 \cdot 10^{13}$ cm^{-3} the ion temperature rise was $\Delta T_i = 0.6$ keV, which corresponded to $\eta_h \sim 4 \cdot 10^{13}$ eV/(kW·cm^3). In Alcator C, an electron heating regime has been obtained. At a density of $n_e \simeq 10^{14}$ cm^{-3} the electron temperature rise was as much as $\Delta T_e \sim 1.2$ keV and the rise in the ion temperature (through the electrons) was $\Delta T_i \sim 0.8$ keV, so that $\eta_h \simeq 2 \cdot 10^{14}$ eV/(kW·cm^3). A major difficulty in these experiments was a rise in the impurity density and a consequent increase in radiation losses. The highest heating efficiency was obtained using carbon limiters coated with silicon carbide. Recently, however, doubts have arisen about the suitability of using carbon limiters, since it was found that there is a reduction in the maximum heating power, apparently caused by an influx of carbon into the vessel.

As to the problems associated with using LHH in large thermonuclear experiments and in a thermonuclear reactor, it should first be noted that the extrapolation from existing results to these systems is not obvious. The role played by scattering of lower hybrid waves on fluctuations in tokamak plasmas has not been fully studied. The operation of antenna systems at high power levels and the effect of the edge plasma on their operation have not been fully clarified. The conditions under which parametric instabilities develop and their effect on the penetration of waves to the plasma center, on ion bombardment of the walls, and on impurity influx are not completely understood. The task of limiting the influx of impurities into reactor-sized machines may turn out to be more complicated than in present-day experiments. The effect of heating on plasma stability and confinement should also be mentioned. Up to now this effect has been unimportant; however, the question of its extent in reactor-sized machines, with their considerably higher powers, dimensions, and plasma confinement times, is still open.

Chapter 5

ION CYCLOTRON HEATING

5.1. THE PHYSICS OF WAVE PROPAGATION FOR $\omega \sim \omega_{Bi}$. PRINCIPAL HEATING SCHEMES

As noted in Section 2.9, at frequencies on the order of the ion cyclotron frequency, the picture of wave propagation in plasmas depends strongly on the ion composition. In plasmas with a single ion species, this picture is fairly simple [51]. The characteristic behavior of the refractive indices over the plasma diameter in this case can be understood from Figs. 5.1 ($\omega \gtrsim \omega_{Bi}$) and 5.2 ($\omega \lesssim \omega_{Bi}$), which illustrate the dependences of the perpendicular refractive indices $N_\perp$ for two waves on R, the distance to the symmetry axis in the equatorial plane of the torus. These dependences can easily be obtained from the general dispersion equation for waves in a cold plasma (2.25) assuming $N_\perp^2 = \text{const}$, which is quite appropriate for a system with weak toroidicity and a low poloidal field. The refractive indices for the fast and slow waves with $N_\perp^2 > 1$ are given by Eqs. (2.102) and (2.104). In accordance with Eq. (2.104), the cutoff points for the slow wave, $N_\perp^2 = 0$, are determined by the condition $\omega = \omega_{pe}(r)$. When $\omega > \omega_{Bi}$, the refractive index of the slow wave in the cold plasma approximation goes to infinity at the lower hybrid resonance point $\epsilon = 0$ (i.e., at $\omega \simeq \sqrt{\omega_{pi}^2 + \omega_{Bi}^2}$). When the thermal motion of the ions is taken into account, in the neighborhood of this point the initial mode undergoes linear conversion into a still slower plasma wave which propagates toward the plasma edge when $\omega > 2\omega_{Bi}$ and into the depth of the plasma when $\omega_{Bi} < \omega < 2\omega_{Bi}$. (See the dashed curve in Fig. 5.1.) In the frequency range of interest to us, on the order of the ion cyclotron resonance, the lower hybrid resonance occurs at an extremely low plasma density ($\omega_{pi} \sim \omega_{Bi}i$). Thus, for $\omega > \omega_{Bi}$ the slow wave and linear conversion into a plasma mode are of no interest for plasma heating if we ignore parasitic effects which may develop in the antenna structure. The Bernstein wave (a strongly slowed-down plasma wave which propagates into the depth of the plasma) may, however, may be used for heating.

When $\omega < \omega_{Bi}$, the lower hybrid resonance is impossible and a slow wave may exist for $N_{\|}^2 > 1$ at plasma densities given approximately by

$$\omega^2 < \omega_{pe}^2 < (N_{\|}^2 - 1)(\omega_{Bi}^2 - \omega^2)|\omega_{Be}|/\omega_{Bi}. \tag{5.1}$$

The upper bound of the accessible densities is determined from the condition for a local Alfvén resonance $\epsilon(r) = N_{\|}^2$, near which linear conversion of the fast and slow waves into one another takes place.

The square of the perpendicular refractive index of the fast wave, $N_{\perp}^2$, is positive when $\omega > \omega_{Bi}$ and $N_{\|}^2 > 1$ for any plasma density above the cutoff given by $(N_{\perp}^2 = 0)$:

$$\omega_{pi}^2 \, (r) / [\omega_{Bi} \, (\omega_{Bi} + \omega)] = N_{\|}^2 - 1. \tag{5.2}$$

Under real experimental conditions, waves with a finite refractive index N_{ϑ} in the poloidal direction, which is given by $N_{\vartheta} = (c/\omega)m/r$ (m is an integer) when the weak toroidal inhomogeneity of the system is neglected, are always excited. In this case the region for propagation of fast waves in a tokamak is found from the condition

$$N_{\perp}^2 = N_{\vartheta}^2 + N_r^2 > N_{\vartheta}^2, \tag{5.3}$$

where $N_{\perp}$ is the perpendicular refractive index of the fast wave given by Eq. (2.102) and N_r is the radial component of the refractive index. The opacity region at the edge ($N_r^2 < 0$), therefore, extends to somewhat higher plasma densities than implied by Eq. (5.2). Of course, Eq. (5.3) also determines the actual boundary of the propagation region for slow waves, if $N_{\perp}^2$ is taken as the value obtained from the dispersion relation for the slow wave (2.104).

An important difference in the propagation behavior for the fast wave with $\omega < \omega_{Bi}$ (Fig. 5.2) compared to the high-frequency case $\omega > \omega_{Bi}$ is related to the existence of a local Alfvén resonance,

$$\epsilon = 1 + N_A^2 \, (1 - \omega^2/\omega_{Bi}^2)^{-1} = N_{\|}^2, \tag{5.4}$$

near which mutual linear conversion of the fast and slow waves occurs and to the presence of two cutoff points for the fast wave $N_{\perp}^2 = 0$. The "edge" cutoff point is given by

$$\omega_{pi} \, (r) = \, (N_{\|}^2 - 1) \, \omega_{Bi} \, (\omega_{Bi} - \omega), \tag{5.5}$$

and the "interior" cutoff, by Eq. (5.2). Thus, there are two opacity regions for the fast wave with $\omega < \omega_{Bi}$: an edge region, which for modes with $m = 0$ is located between the plasma boundary and the point determined by Eq. (5.5), and an interior region, which is located between the points determined by Eqs. (5.2) and (5.4).

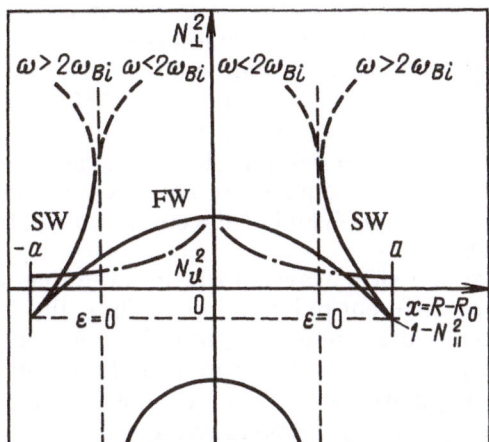

Fig. 5.1. The distribution of the square of the perpendicular refractive index $N_\perp^2$ for fast (FW) and slow waves (SW) in a plasma layer with $\omega > \omega_{Bi}$: —— cold plasma approximation; - - - - - including a finite ion Larmor radius; ------- N_ϑ^2.

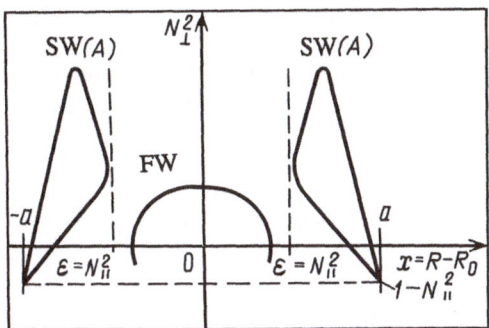

Fig. 5.2. The distribution of $N_\perp^2$ in a plasma layer with $\omega < \omega_{Bi}$.

For $m \neq 0$ these opacity regions are somewhat larger since, according to Eq. (5.3), the actual turning points for the fast wave ($N_r = 0$) are then located at higher plasma densities.

In the low-frequency region $\omega < \omega_{Bi}$ these waves have special names: outside the conversion region the fast wave is known as the fast magnetosonic wave and the slow wave is known as the Alfvén wave.

This picture of wave propagation is valid only when the conditions for the cold plasma approximation are satisfied:

$$\omega^2 / (k_\parallel^2 v_{Te}^2) \gg 1, \quad (\omega - \omega_{Bi})^2 / (k_\parallel^2 v_{Ti}^2) \gg 1, \quad k_\perp^2 v_{Ti}^2 / \omega_{Bi}^2 \ll 1. \tag{5.6}$$

In heating experiments it is easy to realize a situation in which the parallel refractive index $N_\parallel$ is so large that the first of these conditions is violated, even at the edge of the plasma. The conversion and propagation behavior of the fast and slow waves then changes significantly (Fig. 5.3). Thus, for $\omega^2/(k_\parallel^2 v_{Te}^2) \ll 1$ the slow wave propagates only in the dense plasma region and the fast wave, which exists on the edge, is continuously transformed into this slow mode as it moves into the depth of the plasma. This change is a consequence of the way the parallel component of the dielectric tensor ϵ_{zz} depends on $N_\parallel$ and of the singularity in the dispersion relation (2.104) for the slow wave.

Of course, the fast and slow waves differ in more than their radial refractive indices; they also have different polarizations. Thus, the fast wave, as noted in Chapter 2, has an anomalously small electric field component along the external magnetic field. In the plane perpendicular to the external magnetic field this wave generally has an elliptic polarization which becomes circular as $\omega \to \omega_{Bi}$ with a direction of rotation opposite the cyclotron rotation of the ions. In the region with $\epsilon \ll N_\parallel^2$ the slow wave is practically longitudinal ($\mathbf{E} \parallel \mathbf{k}$) for $N_\parallel^2 \gg 1$, while it has a mainly radial electric field in the region with $\epsilon \lesssim N_\parallel^2$.

It follows from this polarization character of the waves that the excitation of fast waves requires antenna systems with predominantly poloidal currents and the excitation of slow waves requires launching systems with electric currents that have a large component along the toroidal magnetic field.

We now consider the case of a plasma with two ion species that have significantly different cyclotron frequencies ω_{B1} and ω_{B2} ($\omega_{B1} > \omega_{B2}$) such that the lighter ions (with ω_{B1}) are present as a minority additive or slight impurity. The transverse component of the dielectric constant ϵ, which determines the principal characteristic points for the fast wave, has the following form in the cold plasma approximation (see Chapter 2):

$$\epsilon = 1 + \omega_{pe}^2/\omega_{Be}^2 - \omega_{p2}^2/(\omega^2 - \omega_{B2}^2) - \omega_{p1}^2/(\omega^2 - \omega_{B1}^2), \tag{5.7}$$

where ω_{p2} and ω_{p1} are the plasma frequencies of the heavy and light ions, respectively. Similarly, adding ions of a second type leads to the appearance of additional terms in the other components of the tensor. Clearly, for a small additive concentration, with $\omega_{p1} \ll \omega_{p2}$, the presence of the additional small terms in the dielectric tensor is unimportant everywhere except in the neighborhood of the cyclotron resonance $\omega \to \omega_{B1}$. For the resonance $\omega = \omega_{B1}$ the components ϵ and g have a pole, so that in this neighborhood these terms cease to be small corrections and one can expect a substantial change in the dispersion of the waves. Thus, when $\omega > \omega_{B2}$ and the resonance $\omega = \omega_{B1}$ is satisfied somewhere inside the plasma, the situation illustrated schematically in Fig. 5.4 occurs. The function $N_\perp^2(x)$ ($x = R - R_0$) shown there transforms into the corresponding curve in Fig. 5.1 when $\omega_{p1} = 0$. It is clear that when $\omega_{p1} \neq 0$ there is yet another existence region for the slow wave, owing to linear conversion from the fast wave, as well as a resonance point for the slow wave $N_\perp^2 \to \infty$ (the ion–ion hybrid resonance). When thermal motion is taken into account in the neighborhood of this resonance, linear conversion into a plasma wave takes place as usual. This is shown in Fig. 5.4 by a dashed curve. Between the cutoff points ($N_\perp^2 = 0$) and the fast–slow wave conversion points ($\epsilon = N_\parallel^2$) there is an opacity region whose depth may vary over wide limits, depending on the concentration of the minority additive and on the size of the system.

This form of the dispersion curves $N_\perp^2(x)$ makes it possible to represent the propagation behavior of the waves in the equatorial plane of the system as follows: a fast wave launched from the high magnetic field side (from the inner edge of the torus) tunnels through the edge opacity barrier and propagates along the x axis to the point $x = x_c$ ($\epsilon = N_\parallel^2$). Near this point it is partially converted into the slow mode and partially penetrates the opacity barrier in the region $x > x_0$. The slow wave resulting from this conversion propagates inward to the ion–ion hybrid resonance point x_{ii} [$\epsilon(x_{ii}) = 0$] and is completely converted into a plasma mode near that point. It makes sense that part of the energy of the incident wave should be absorbed as it propagates, but we shall discuss this later and only consider wave conversion for now. The efficiency with which the fast wave is converted to the slow wave depends strongly on the width of the opacity barrier $x_0 - x_c$. For a thick barrier, when the characteristic wavelength of the wave as it approaches the conversion region is much smaller than the barrier thickness, almost all of the energy of the incident wave is converted and for a thin barrier, only a small fraction of the energy is converted, while the remainder penetrates into the region beyond (to the right of) the barrier.

When a fast wave is incident on the opacity barrier from the outside, part of its energy is reflected, part penetrates through the barrier in the form of a fast mode, and part is converted into a slow wave. At an opaque barrier [$(\omega/c) \int_{x_0}^{x_c} | N_\perp | dx \gg 1$] the reflection coefficient is close to unity and the conversion efficiency is exponentially small. As the width of the barrier approaches zero, both the reflection coefficient and the conversion efficiency go to zero, so that the conversion is greatest for some intermediate barrier thickness.

The width of the opacity barrier for a given geometry is proportional to the relative impurity density. For purposes of illustration we now give formulas [51] for the distances between the characteristic points of the refractive index in a deuterium plasma with a small admixture of hydrogen (Fig. 5.4):

the distance between the cyclotron resonance point for hydrogen $x = x_r$ and the cutoff point $x = x_0$

$$x_r - x_0 \simeq (1/2)(n_H/n_D)(1 + N_\parallel^2/N_A^2)^{-1} R_0; \tag{5.8}$$

the distance between $x = x_r$ and the ion–ion hybrid resonance point x_{ii}

$$x_r - x_{ii} = (3/4)(n_H/n_D)R_0; \tag{5.9}$$

the width of the opacity region

$$x_0 - x_c \simeq \frac{1}{4}\frac{n_H}{n_D}R_0; \tag{5.10}$$

the displacement of the conversion point for the fast and slow waves relative to the point $x = x_c$ where $\epsilon = N_\parallel^2$,

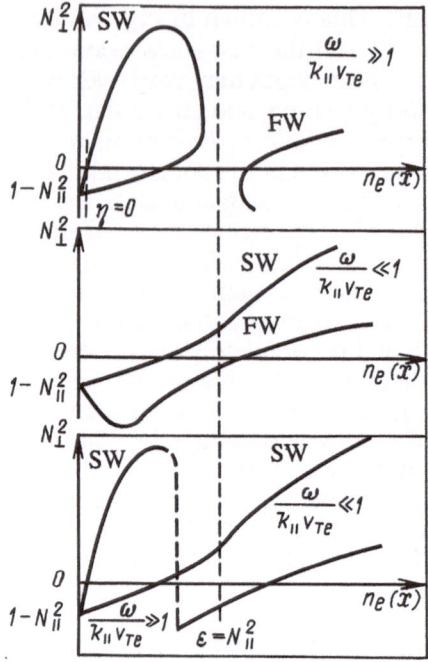

Fig. 5.3. The dependence of $N_\perp{}^2$ on the plasma density for different ratios of the longitudinal phase velocity of the wave to the electron thermal speed.

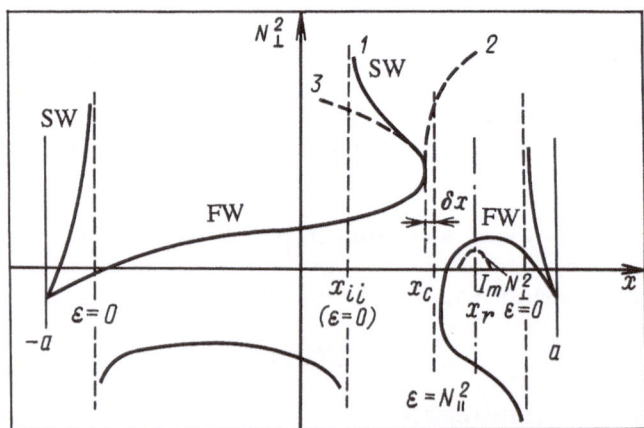

Fig. 5.4. The distribution of $N_\perp{}^2$ over the plasma cross section when a small amount of light impurity ions are present: 1) cold plasma approximation; 2) $\beta_{eff} > 0$; 3) $\beta_{eff} < 0$.

$$\delta x = 3 \left(\frac{m_e}{m_i} \right)^{1/2} \frac{n_H}{n_D} \frac{N_\parallel}{N_A} R_0, \tag{5.11}$$

and the width of the propagation region for the slow wave (in the cold plasma approximation)

$$x_c - x_{ii} = \frac{9}{4} \frac{n_H}{n_D} \frac{N_\parallel^2}{N_A^2} R_0. \tag{5.12}$$

In Eqs. (5.8)–(5.12) n_D is the density of deuterium ions, n_H is the density of hydrogen ions, $n_H/n_D \ll 1$, and $N_A^2 = \omega_{pD}^2/\omega_{BD}^2$ is evaluated at $x = x_r$. In deriving these formulas it was assumed for simplicity that $N_\parallel^2 \ll N_A^2$. Without added hydrogen, the refractive index for the fast wave in the region of interest to us, $\omega = 2\omega_{BD}$, is given by

$$N_\perp^2 = (N_A^2 + N_\parallel^2 - 1)(N_A^2 - 3N_\parallel^2 + 3)(N_A^2 + 3N_\parallel^2)^{-1} \approx N_A^2, \tag{5.13}$$

where it is noted that $N_A^2 \gg 1$. This expression can be used to estimate the depth of the opacity region at the plasma edge. It is easy to show that for a parabolic density profile, when $N_A^2(x) = N_{A0}^2(1 - x^2/a^2)$, the condition that the barrier be transparent, $(\omega/c)\int |N| \, dx \ll 1$, yields the inequality

$$N_\parallel^3 \ll N_{A0}^2 \, c/\omega a.$$

In order to get an idea of how the waves propagate outside the equatorial plane and throughout the entire volume of a tokamak, Fig. 5.5 shows the locations of the characteristic curves and opacity regions $N_\perp^2 < 0$ in the minor cross section of a plasma column. As noted previously, when $N_\vartheta \neq 0$ the actual regions in which the waves propagate differ from the regions with $N_\perp^2 > 0$. Here, as can be seen from Fig. 5.5, the plasma column is highly nonuniform for the waves, both along its minor radius and in the poloidal direction (because of the ion–ion hybrid resonance and the fast wave cutoff inside the plasma), so that the equation $N_\vartheta = (c/\omega)m/r$ is not satisfied even approximately and the difficult problems of finding the orientation of the vector $N_\perp = k_\perp c/\omega$ and the localization of the wave fields in the plasma column are mostly still unsolved. Qualitatively the situation is somewhat simpler when the interior opacity region is fairly wide and there is little tunneling of the waves through it. Then the region on the right (outside) with $N_\perp^2 > 0$ forms a unique toroidal resonator in which an antenna structure located on the outside of the torus can excite a certain number of eigenmodes. Almost no field is excited in the inner transparency region by such an antenna. Such a field can be excited by an antenna located on the inside of the torus and consists of a beam of quasiclassical waves travelling toward the layer with $\epsilon = N_\parallel^2$ which is almost completely converted there (for a wide opacity barrier). For a very narrow barrier the conversion efficiency is low and a toroidal resonator again develops, but it now occupies the entire plasma region where $N_r^2 > 0$. The inner opacity region then only slightly perturbs the eigenmodes of the resonator and mode conversion lowers the Q of the resonator.

When there is no cyclotron resonance for the ions of a minority additive, the dispersion of the waves changes little and the picture given in Fig. 5.2 remains the same when a light impurity (more precisely, one with a high cyclotron frequency) is introduced into the plasma. It does change when heavier ions for which the cyclotron resonance condition $\omega = \omega_{B2}$ is satisfied are introduced into the plasma. These changes are similar in character to those examined previously (Fig. 5.6).

We now consider the changes in the propagation behavior owing to spatial dispersion (thermal motion of the particles). We have already discussed the effect of parallel spatial dispersion on the localization of the region where slow waves exist in a plasma with a single ion species. We now consider its role in a plasma with two ion species. It follows from Eq. (5.7) that in the cold plasma approximation the addition of an arbitrarily low density of impurities will lead to a substantial change in the form of $N_\perp^2(x)$ because of the pole at the cyclotron resonance. When the parallel motion of the ions is taken into account, however, this is not so, since it limits the resonant contribution of these ions to ϵ to an amount $\omega_{pi}'^2/(\omega_{Bi}'k_\parallel v_{Ti}')$ (cf. Chapter 2). (Here the primes denote quantities that apply to the additive.) If this contribution is considerably smaller than the contribution to ϵ from the primary ions, i.e., if

$$\omega_{pi}'^2 \mid \omega_{Bi}^2 - \omega_{Bi}'^2 \mid /(\omega_{pi}^2 \, \omega_{Bi}' \, k_\parallel v_{Ti}') \ll 1, \tag{5.14}$$

then the real part of the refractive index varies little even in the neighborhood of the resonance $\omega = \omega_{Bi}'$ and the presence of a small amount of resonant ions mainly leads to wave damping. The above picture of fast and slow wave conversion, ion–ion hybrid resonances, etc., is, therefore, valid when the opposite inequality holds. That condition can be written in terms of the impurity density n_i' as

$$n_i' \gg n_{\mathrm{ch}} = N_\parallel (v_{Ti}'/c) n_e. \tag{5.15}$$

Transverse spatial dispersion (the finite Larmor radius of the ions) has a strong effect on wave conversion. Thus, in a plasma with a single ion species it leads to conversion of the fast wave into a plasma (ion cyclotron) wave in the neighborhood of the resonance at the second harmonic of the ion cyclotron frequency $\omega = 2\omega_{Bi}$. This can also be seen from the structure of the finite temperature correction for the perpendicular component of the dielectric tensor

$$\epsilon_{xx} = \epsilon - \frac{3}{2} \frac{\omega_{pi}^2 \, k_\perp^2 \, v_{Ti}^2}{(\omega^2 - \omega_{Bi}^2)(\omega^2 - 4\omega_{Bi}^2)}, \tag{5.16}$$

as written in the approximation of weak spatial dispersion. In fact, this correction has a resonance at $\omega = \omega_{Bi}$ and in this sense is analogous to the correction to ϵ in Eq. (5.7) for a minority additive. Far from the resonance this correction is small provided

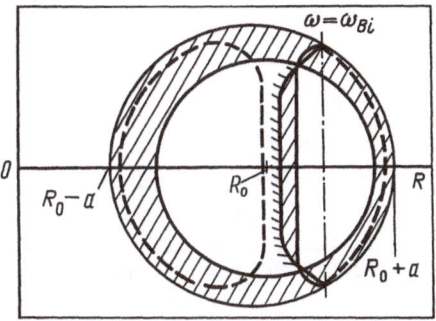

Fig. 5.5. The poloidal locations of the resonances and cutoffs in a plasma: —— fast wave cutoffs ($N_\perp^2 = 0$); - - - - hybrid resonance ($N_\perp^2 \to \infty$); ⫽⫽ conversion of the fast and slow modes; and ····· cyclotron resonance of the additive species $\omega = \omega_{Bi}$. The opacity regions ($N_\perp^2 < 0$) are shaded.

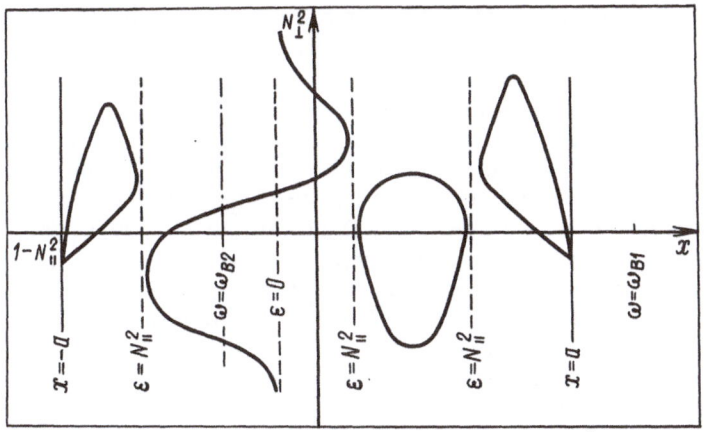

Fig. 5.6. The distribution of $N_\perp^2$ in the plasma cross section when $\omega \sim \omega_{B2}$ and $\omega < \omega_{B1}$.

$$N_A^2 \, v_{Ti}^2 / c^2 = v_{Ti}^2 / v_A^2 \ll 1, \tag{5.17}$$

a condition which is easily satisfied in the plasma.

As usual, including spatial dispersion leads to the appearance of an additional branch of strongly retarded parallel waves. In the region where the condition $N_\perp^2 \gg N_\parallel^2$ is satisfied with a wide margin for this branch, the dispersion for these waves will obey the approximate equation (see Chapter 2)

$$\epsilon_{xx} = 0. \tag{5.18}$$

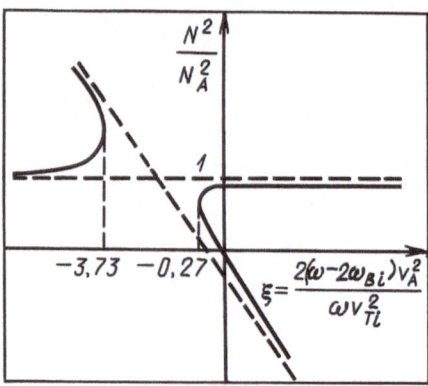

Fig. 5.7. The dispersion of waves near the second harmonic of the ion cyclotron resonance frequency [51].

Then, using Eq. (5.16) it is evident that the refractive index for this wave decreases as $\omega \to 2\omega_{Bi}$. When it approaches the refractive index for the fast wave $N_\perp^2 \approx N_A^2$, the assumption that the field of the plasma wave is longitudinal breaks down and this mode begins to interact with the fast wave.

The function $N_\perp^2(\xi)$ in the neighborhood of the resonance $\omega = 2\omega_{Bi}$ for small values of $N_\parallel$ which satisfy the condition $N_\parallel \ll N_A v_{Ti}/v_A$ is shown in Fig. 5.7. Here the conversion efficiency is determined by the "depth" of the opacity region lying between the points $\xi = -3.73$ and $\xi = -0.27$. When $N_\parallel \sim N_A v_{Ti}/v_A$, the parallel spatial dispersion must be taken into account in the conversion region. Then the refractive indices of the waves will be complex (i.e., the waves will be damped) and the conversion will be more complicated. In the case $N_\parallel \gg N_A v_{Ti}/v_A$, parallel dispersion limits the finite temperature correction to Eq. (5.16) to such a low level that the refractive index for the fast wave changes little as $\omega \to 2\omega_{Bi}$ and conversion into a plasma wave does not take place, although, of course, finite damping of the wave still occurs in the region $|\omega - 2\omega_{Bi}| \leq N_\parallel \omega_{Bi} v_{Ti}/c$.

We note that because of the rapid spatial variation of both the refractive index of the waves and the parameter $(\omega - 2\omega_{Bi})^2 (k_\parallel v_{Ti})^{-2}$ in the resonance region, the conditions for applicability of geometric optics to this case are rather strict. More rigorous theoretical treatments of phenomena near the resonance do exist [51].

Perpendicular spatial dispersion also has a strong effect on the conversion of fast and slow waves in a plasma with two ion species, especially in a deuterium plasma with a small admixture of hydrogen. In such a plasma the locations of the cyclotron resonance for hydrogen and the resonance at the second harmonic of the cyclotron frequency for deuterium are practically the same, so that the correction to ϵ_{xx} for the hydrogen ions and the dispersion correction for the main plasma have poles at the same point.

If Eq. (5.17) is satisfied, the spatial dispersion has little effect on the refractive index of the fast wave far from the conversion region. Because of the large value $N_\perp^2 \approx |g| \sqrt{-\pi/\epsilon} \approx (N_A^3/6N_\parallel) \sqrt{m_i/m_e}$ of the refractive index, spatial dispersion can be properly neglected in the conversion region only when the much more restrictive condition

$$6 \frac{n_H}{n_D} \frac{v_A^2}{v_{TD}^2} \frac{N_\parallel}{N_A} \sqrt{\frac{m_e}{m_i}} \gg 1, \tag{5.19}$$

which is rather unrealistic for tokamaks, is satisfied. Consequently, the conversion behavior is fairly complicated. As the conversion region is approached, while the refractive index of the fast wave is still not large ($N_\perp^2 \sim N_A^2$), spatial dispersion is unimportant and the resonant increase in $N_\perp^2$ is caused mainly by the small amount of added hydrogen. However, in the conversion region it is already very important. In particular, the conversion point is displaced [51] relative to the point $x = x_c$ by

$$\delta_x = R_0 \, [3v_{TD}^2 n_H/(4v_A^2 \, n_D)]^{1/2}, \tag{5.20}$$

and the refractive index at this point takes the form

$$N_\perp^2 \simeq N_A^2 \, [n_H v_A^2 / (6 n_D \, v_{TD}^2)]^{1/2}.$$

The shape of the $N_\perp^2(x)$ curve is also modified considerably for the slow wave, which is now continuously being transformed into the plasma wave described by Eq. (5.18). (See the dashed curve 3 in Fig. 5.4.) The cold plasma approximation is essentially unsuitable for the slow waves.

In the general case of an arbitrary relation between ω_{B1} and ω_{B2} and between the densities of the different ions, the behavior of $N_\perp^2(x)$ in the fast and slow wave conversion region is determined by the value and sign of the parameter

$$\beta_{eff} = \frac{1}{2c^2} \sum_i \frac{3 v_{Ti}^2 \, \omega_{pi}^2 \, \omega^2}{(\omega^2 - \omega_{Bi}^2)(\omega^2 - 4\omega_{Bi}^2)}, \tag{5.21}$$

where the sum is taken over all ion species present in the plasma. The variation in the refractive index for the case $\beta_{eff} < 0$, which is realized in deuterium–hydrogen plasmas, is shown by the dashed curve 3 in Fig. 5.4. The dashed curve 2 corresponds to $\beta_{eff} > 0$. Far from the conversion region, when Eq. (5.17) is satisfied and $\omega \neq 2\omega_{Bi}$, transverse spatial dispersion only leads to small corrections in the refractive index for the fast wave. It completely determines the variation of the refractive index in the conversion region when

$$|\beta_{eff}| \frac{\omega_{pe}^2}{\omega^2} \approx \frac{v_{Ti}^2}{v_A^2} N_A^2 \frac{|\omega_{Be}|}{\omega_{Bi}} \gg N_\parallel^2. \tag{5.22}$$

The refractive index of the slow plasma wave that results from the conversion is then given by the ratio

$$N_\perp^2 = (\epsilon - N_\parallel^2)/\beta_{eff}, \qquad (5.23)$$

which differs substantially from the dispersion equation for the slow wave in a cold plasma (2.104), but is easily derived from it by replacing ϵ by $-\beta_{eff}N_\perp^2$. Equation (5.23) can also easily be derived by making this substitution in the dispersion relation for the fast wave (2.102).

It should be kept in mind that spatial dispersion can be taken into account using the finite temperature correction (5.21) [as, for example, in Eq. (5.16) for a plasma with a single ion species] only near the conversion region, where the refractive index is still not very large. Equation (5.16) represents the lower-order terms in the series expansion of a more accurate expression in terms of the parameter $k_\perp^2 v_{Ti}^2/\omega_{Bi}^2$, which is assumed to be small. With increasing distance from the conversion point the refractive index of the plasma wave increases rapidly, this expansion is no longer valid, and more accurate formulas must be used. A simplification results from the fact that far from the conversion region the plasma waves can, with great accuracy, be regarded as longitudinal and can be described by the equation

$$k_\alpha \epsilon_{\alpha\beta} k_\beta = 0, \qquad (5.24)$$

where $\epsilon_{\alpha\beta}$ is the dielectric tensor for a hot plasma.

The efficiency of linear wave absorption in hot plasmas at frequencies comparable to the ion cyclotron frequency has been under intensive discussion in the literature. It has been examined in detail for the conditions of ion cyclotron plasma heating in a review by Longinov and Stepanov [51]. Here we shall list some formulas from that review which are useful for estimating the relative contributions of various wave absorption mechanisms.

As at higher frequencies, collisional absorption at $\omega \sim \omega_{Bi}$ is fairly weak, and in most cases of practical interest it can be neglected. The primary mechanisms are collisionless: electron Landau damping (Cerenkov damping) and cyclotron absorption. We shall examine these briefly as applied to the situations examined above. We begin with Landau damping in a plasma with one ion species. The damping coefficient for fast waves in this sort of plasma is fairly low, since the fast wave has a small component of the electric field along the constant magnetic field and it is this component which is responsible for energy absorption by electrons through Landau damping. For a dense plasma with $N_\perp^2 \gg N_\parallel^2$ it has the form

$$\operatorname{Im} N_\perp \simeq N_\perp \frac{\sqrt{\pi}}{2} \frac{\omega_{pe}^2}{\omega_{Be}^2} \frac{T_e}{m_e c^2} z_e \exp(-z_e^2), \qquad (5.25)$$

where $z_e = \omega/(k_\parallel v_{Te})$. From this it is easy to obtain an estimate, when $N_\perp \gg N_\vartheta$, for the optical thickness of the plasma, $\Gamma = (\omega/c)\int_0^a \operatorname{Im} N_r dr$, which characterizes the absorption efficiency in a single pass through the plasma column:

$$\Gamma = \overline{z_e \exp(-z_e^2)} N_A \frac{\omega a}{c} \frac{\omega_{pe}^2}{\omega_{Be}^2} \frac{T_e}{m_e c^2}, \qquad (5.26)$$

where the bar denotes the average along the ray. For tokamak reactor conditions Eq. (5.26) gives $\Gamma \approx 0.5$, and for a tokamak with the parameters of T-10, $\Gamma = 10^{-2}$–10^{-3}.

The efficiency of Landau damping of slow waves is considerably higher:

$$\text{Im}\, N_\perp \simeq N_\perp \sqrt{\pi}\, z_e^3 \exp(-z_e^2). \qquad (5.27)$$

Evidently, slow waves are damped weakly only when the following condition is satisfied with a wide margin: $z_e \gg 1\left(z_e^2 > 3 + \ln\left(\frac{\omega a}{c} N_\perp\right) + \frac{3}{2}\ln\ln\left(\frac{\omega a}{c} N_\perp\right)\right)$.

As noted before, the fast wave can be damped in the neighborhood of the resonance at the second harmonic of the ion cyclotron frequency.

The optical thickness of the plasma near the resonance when $N_\parallel 2 \gg v_{Ti} N_A^2/c$ can be estimated using [51]

$$\Gamma \simeq (\text{Im}\, N_r)_{max} \frac{\omega \Delta R}{c} \simeq \frac{4\pi n_i\, T_i\, N_A}{B_0^2}\, \frac{\omega R_0}{c}. \qquad (5.28)$$

For tokamak reactor conditions, the wave damping given by Eq. (5.28) is fairly strong ($\Gamma > 1$) and for tokamaks the size of T-10 it is relatively weak ($\Gamma \leq 0.1$).

When $N_\parallel < v_{Ti} N_A^2/c$, as we have noted, wave conversion plays an important role (Fig. 5.7). A fast wave incident on the resonance layer from the high magnetic field side is almost completely converted into a plasma mode in a thick opacity barrier and does not reach the $\omega = 2\omega_{Bi}$ resonance. When it is incident from the low field side, after it passes through the resonance layer a fast wave is converted into an ion cyclotron wave whose refractive index decreases as it approaches the point where $\omega = 2\omega_{Bi}$. Calculations show that the optical thicknesses of the resonance layer for the fast (Γ_f) and ion cyclotron (Γ_i) waves are given by

$$\Gamma_f \sim 0.1 \frac{\omega R_0}{c}\, \frac{N_\parallel^2}{N_A^2},$$
$$\Gamma_i \sim \frac{\omega R_0}{c} N_\parallel^{3/2} \sqrt{v_{Ti}/c}. \qquad (5.29)$$

The imaginary parts of the components of ϵ and g are large near the cyclotron resonance $\omega = \omega_{Bi}$. As discussed in detail in Chapter 2, however, in a plasma with one ion species the fast wave is absorbed weakly at the cyclotron resonance because of the small field component rotating in the direction of the cyclotron orbits of the ions. For sufficiently large $N_\parallel$ the order of magnitude of the optical thickness of the cyclotron layer of the plasma can be evaluated using the formula

$$\Gamma \simeq \frac{1}{5}\, \frac{N_\parallel^2\, v_{Ti}^2}{N_A\, v_A^2}\, \frac{\omega R_0}{c}. \qquad (5.30)$$

Wave absorption in the neighborhood of the cyclotron resonance of an impurity is entirely different. When condition (5.14) is fulfilled, a small amount of resonant impurity ions has little effect on the refractive index of the wave or on its polarity. But since the wave has a significant field component rotating in the ion direction far from the cyclotron resonance of the primary ions, the impurity ions may be efficiently heated by this field near their cyclotron resonance and absorb the energy of the wave. As a result, the optical thickness of the layer for $n_i' \ll n_{ch}$ is given by

$$\Gamma \simeq (n_i'/n_e)N_A \, \omega R_0/c. \tag{5.31}$$

This formula shows that absorption of the wave increases with the impurity density n_i'. This result, however, is valid only for $n_i' \ll n_{ch}$. When $n_i' \gg n_{ch}$, the impurity already has a significant effect on the polarization of the wave while it reduces the resonant component of the field. Thus, the absorption efficiency also decreases:

$$\Gamma \sim N_A N_\parallel^2 \frac{v_{Ti}'^2}{c^2} \frac{n_e}{n_i'} \frac{\omega R_0}{c}. \tag{5.32}$$

From this it follows that the optical thickness of the cyclotron layer for the impurity ions reaches a maximum which depends on the density n_i' when $n_i' \sim n_{ch}$. The order of magnitude of this maximum can be estimated using Eq. (5.31) with $n_i' \sim n_{ch}$. It turns out that for the conditions in tokamak reactors and smaller tokamaks $\Gamma > 1$ at the maximum.

Waves can, of course, also be absorbed near the resonance at the second harmonic of the cyclotron frequency of the impurity ions ($\omega = 2\omega_{Bi}'$). This case is interesting [51] in connection with the fact that tokamak plasmas often contain highly charged impurity ions. The cyclotron frequency of highly ionized heavy ions, $\omega_{Bi}' = eZ'B_0/(m_H A'c)$ (with charge Z' and atomic mass A'), is close to (usually somewhat lower than) the cyclotron frequency of deuterium ions. As a result, in a deuterium–hydrogen plasma it may happen that the resonance $\omega = 2\omega_{Bi}'$ is localized near the point at which the fast wave is converted into the slow mode. Then, because the waves are strongly slowed down, they are absorbed much more rapidly. Calculations show that in most tokamaks the optical thickness of the resonant region in the plasma for this case can be on the order of unity even when the impurity density is extremely low, $n_i'T_i'/(n_D T_D) \sim 10^{-3}$.

As the conversion point is approached, damping of the fast wave on electrons increases sharply. This can be seen from the exact formula for the damping coefficient:

$$\text{Im}\, N_r \simeq \frac{\omega^2}{\omega_{pe}^2} N_r \frac{\epsilon g^2 \sqrt{\pi} \, z_e^3}{(\epsilon - N_\parallel^2)^2} \exp(-Z_e^2). \tag{5.33}$$

At the conversion point ($\epsilon = N_\parallel^2$) itself, $\text{Im}\, N_\perp$ is given by

$$\text{Im}\, N_r \sim N_r \, [(\sqrt{\pi}/2) z_e^3 \exp(-z_e^2)]^{1/2}. \tag{5.34}$$

It follows from Eqs. (5.33) and (5.34) that when Z_e is not very large, the fast wave is almost completely absorbed by electrons in the conversion region.

The damping of the slow plasma waves formed by conversion of the fast wave depends on $N_\parallel$. For small $N_\parallel$, cyclotron absorption at the resonances $\omega = s\omega_{Bi}$, where $s = 1, 2, ...$, predominates, while for $N_\parallel^2 \gg 1$, electron Landau damping usually predominates in large tokamaks. Nonlinear processes may also make an important contribution to the absorption of plasma waves.

A number of schemes for heating plasmas in toroidal systems have been proposed on the basis of the wave propagation and absorption behavior in the ion cyclotron frequency range described above. Many of them have been tested experimentally. We now discuss the most promising of these schemes [51].

Heating of "simple" plasmas by fast magnetosonic waves. The preceding discussion shows that the plasma in a reactor tokamak can be heated by exciting fast waves without using damping on impurities or conversion into the slow mode near the ion–ion hybrid resonance. Efficient absorption of the waves is then ensured by Landau damping or by the resonance at the second harmonic of the ion cyclotron frequency. The waves can be excited by antenna structures located on the low field side of the torus.

In medium-sized tokamaks (T-10 or smaller) the damping of fast magnetosonic waves is weaker, so that the toroidal plasma column forms a high-quality resonator whose eigenfrequencies depend on the plasma parameters, primarily the density and its profile. Under these conditions it may be necessary to maintain the resonator during the discharge by, for example, changing the source frequency, in order to ensure optimum coupling of the antenna to the plasma. Tracking the mode at high heating powers is an extremely complicated technical problem. In large machines, however, the neighboring modes are extremely close to one another and will overlap even with weak damping. Thus, regimes may occur in which the antenna system excites a large number of eigenmodes in the plasma column and the matching conditions will depend only weakly on the density and other discharge parameters. Nevertheless, weak damping of the waves creates certain difficulties in experiments using this heating scheme because parasitic absorption of the rf power (at the vessel walls, in the antenna system, etc.) must be much lower than the absorption in the plasma.

Excitation of fast waves in plasmas with a small minority of resonant ions. We have noted that the damping of fast waves increases greatly when light ions, for which the cyclotron resonance condition is met inside the plasma column, are added to the plasma. Thus, a small amount of hydrogen or of the light isotope of helium (^3He) can be added to a deuterium plasma. When condition (5.14) is satisfied, this minority additive has little effect on the conditions for wave propagation, in particular, on the resonance frequencies of the eigenmodes, but it does determine how they are damped. Under these conditions the rf energy is initially absorbed by the minority ions and is then transferred to the bulk plasma through collisions. During heating at high powers the minority ions are strongly "overheated" and their distribution functions are highly distorted, i.e., fast particles appear. In small tokamaks these particles are poorly confined. They may bombard the vessel walls and antenna structure, causing an increased influx of (heavy) impurities into the discharge.

The energy gained by the minority ions during heating obviously decreases as their density is raised. It is, therefore, desirable that heating be done at the maximum possible minority species density.

Plasma heating under ion–ion hybrid resonance conditions. If the light impurity density is raised to a level $n_i' \gg n_{ch}$, then the ion–ion hybrid resonance and linear conversion of the fast wave into the slow mode have a significant effect on the wave behavior. The details of this picture depend strongly on the specific conditions: the system dimensions, the plasma temperature, $N_\parallel$, etc. Thus, in large machines waves launched from the inner and outer edges of the vessel are absorbed in completely different ways. When waves are launched from the inner edge (i.e., from the high toroidal magnetic field side), fast waves are completely converted into slow waves which are mostly absorbed by the electrons. In this situation a high-quality toroidal resonator is formed in the outer part of the plasma column and an outer antenna will excite the eigenmodes of this resonator.

In small tokamaks the inner opacity region may be shallow, so that the difference in the absorption of waves launched from opposite sides of the torus is not so great.

Plasma heating by slow and ion cyclotron waves. As we have noted, when $\omega_{Bi} < \omega < 2\omega_{Bi}$ the plasma waves which are produced by conversion of the slow wave near the lower hybrid resonance at the plasma edge will propagate into the depth of the plasma. These waves can also be used for heating. Under actual conditions the plasma density near the antenna structure can exceed the value corresponding to the hybrid resonance. Then a specially constructed antenna can excite plasma (ion cyclotron) waves directly. The damping of these waves depends on $N_\parallel$. For small $N_\parallel$ an ion cyclotron wave launched from the outside of the torus will propagate in the direction of increasing magnetic field and may reach the cyclotron resonance $\omega \approx \omega_{Bi}$, near which its energy is completely absorbed by the ions. As $N_\parallel$ increases, the contribution of electron Landau damping of absorption of the wave increases.

Other schemes for plasma heating using slow waves are possible in principle.

The physical picture of heating presented above is, of course, qualitative and extremely approximate. First of all, it does not include a whole series of factors which play an important role under actual experimental conditions. In addition, it provides little information on the energy deposition profile in the plasma column and on ways to optimize it. The notable success of heating experiments in recent years has led to great efforts toward improving our understanding of the physics of heating. Research has developed in two areas. First, improved analytic and numerical models have been developed for obtaining a deeper understanding of the excitation of waves by antennas, the physics of wave conversion and absorption, and the evolution of the energy distribution functions of the plasma particles during heating. Second, attempts have been made to create numerical models of heating which can be used to calculate energy deposition profiles and follow the evolution of the parameters of a plasma when rf power is applied to it.

Here we shall give only a brief description of the main achievements in these areas. More detailed discussions can be found in the reviews by Longinov and Stepanov [51] and Vdovin [159] which contain extensive references.

One of the central topics of recent research is antenna systems. Several three-dimensional models of antennas have been developed [158, 159]. They share the following simplifying assumptions: a plane geometry is considered; it is assumed that there is no plasma in the antenna region; the electrostatic shield that is required to suppress excitation of slow waves is replaced by a plane with anisotropic conductivity; the toroidal inhomogeneity of the magnetic field is neglected; the antenna windings are replaced by current layers with a uniform current distribution inside each; and in calculating the impedance of the plasma boundary, only those waves which penetrate into the depth of the plasma are taken into account.

The $k_\parallel$ distribution of the emitted power plays an important role in the formation of the energy deposition profile. Waves with small $k_\parallel$ are not absorbed efficiently enough inside the plasma, so it is desirable to reduce the spectral distribution $P(k_y, k_\parallel)$ as $k_\parallel \to 0$. Antennas consisting of several windings powered in opposite phase [158, 159] have been proposed for this purpose and have been modelled numerically. Numerical simulations [159] show that an antenna whose windings are embedded in special hollows in the vessel walls can also be extremely efficient.

Important progress has been made in understanding the physics of wave conversion and absorption under real experimental conditions [51, 158]. This progress has been to a great extent related to the development of analytical models which, despite their being greatly idealized, do take more realistic account of the inhomogeneities in the plasma and magnetic field structure. Papers have been published recently which contain numerical solutions of the complete wave equations [160–162]. This type of calculation has been done in a plane geometry with certain simplifying assumptions [160, 161]. Another paper [162] reports the first numerical solution of the wave equations and a determination of the energy deposition profile in a real tokamak geometry with a realistic model for the antenna structure.

A crucial factor in ion cyclotron heating (ICH) of plasmas is the evolution of the ion velocity distribution function as they interact with the rf power (see Section 2.12). Up to now, simplified models in which the structure of the magnetic field in toroidal systems was treated rather approximately [e.g., Eq. (2.162)] have usually been employed. The properties of the solution of Eq. (2.162) as applied to ICH of a minority species have been studied analytically by Stix [56]. His results reduce to the following: the form of the distribution function of the resonant ions is determined by a single parameter ξ_{QL}, which is defined as

$$\xi_{QL} = \frac{m_r \langle p \rangle \sqrt{2T_e}}{8\sqrt{\pi m_e}\, n_e\, n_r\, Z_r^2\, e^4 \Lambda},$$

where m_r, Z_r, and n_r are the mass, charge, and density of the resonant ions, Λ is the Coulomb logarithm, and $<p>$ is the power absorbed by the resonant ions per unit volume, averaged over a magnetic surface.

The ion distribution function $F_r(v)$ can be regarded as approximately isotropic for energies

$$mv^2/2 < \mathcal{E}_c = 14.8\, T_e A_r\, [Z_i T_i/(A_i T_e)]^{2/3}, \tag{5.35}$$

where Z_i, A_i, and T_i are the charge, mass number, and temperature of the background ions and A_r is the mass number of the resonant ions. For low energies,

$$mv^2/2 < \mathcal{E}_c\,[1 + \xi_{QL}\,v_{Ti}/(Z_i v_{Te})]\,(1 + \xi_{QL})^{-1}, \tag{5.36}$$

the distribution function in the steady state is close to Maxwellian with an effective temperature

$$T_{\text{eff}} \simeq T_i\left[1 + \frac{v_{Ti}\,\xi_{QL}\,(\xi_{QL}\,T_e + T_e - T_i)}{Z_i\,v_{Te}\,T_e\,(1 + \xi_{QL})}\right]. \tag{5.37}$$

It is also close to Maxwellian when the opposite inequality holds in Eq. (5.36), but in this case when $\xi_{QL} \gg 1$, T_{eff} is considerably higher:

$$T_{\text{eff}} = T_e\,(1 + \xi_{QL}). \tag{5.38}$$

When $mv^2/2 > \mathcal{E}_c$ the ion velocity distribution is highly anisotropic, but the distribution function for the perpendicular velocity $F_r(v_\perp) = \int_{-\infty}^{\infty} F_r(\mathbf{v})\,dv_z$ can be approximated by a Maxwellian function with a temperature $T_e[1 + (3/2)\xi_{QL}]$.

Further progress in the quasilinear theory of resonant ion heating has long been associated with more accurate numerical modelling of the quasilinear diffusion coefficient and collision integral in Eq. (2.162) [163, 164]. However, attempts have recently [165] been made to model the ICH process in toroidal systems including the actual structure of the magnetic field. The motion of the ions along their drift trajectories has been calculated and their acceleration during each pass through the cyclotron resonance zone is taken into account. It has been shown that above a very low threshold level for the rf power, the ions accumulate energy stochastically. It is interesting to note that according to these calculations ion heating is accompanied by additional diffusion with a diffusion coefficient

$$D_{\text{eff}} \sim P/(B_{0\varphi}^2 B_{0\vartheta}^2), \tag{5.39}$$

where $B_{0\vartheta}$ is the poloidal component of the magnetic field B_0. In experiments on the PLT tokamak this diffusion becomes comparable to the collisional neoclassical diffusion at an input power level of 2 MW.

We now consider the development of complete numerical models of heating which make it possible to calculate energy deposition profiles. Until very recently these models were based, as are those for higher frequency bands, on calculations of ray trajectories and wave damping along them [158]. In the ion cyclotron resonance frequency range, however, the application of ray trajectories is complicated by two factors. First, the dimensions of the antenna system are usually smaller than the vacuum wavelength of the waves, so that diffraction effects may be important. Second, inside the plasma there is an opacity region through which electromagnetic energy tunnels in a fashion not described by geometric optics and this tunneling plays an important role in the heating process. Thus, calculations of a ray trajectory must be stopped upon approaching the opacity region, the field structure

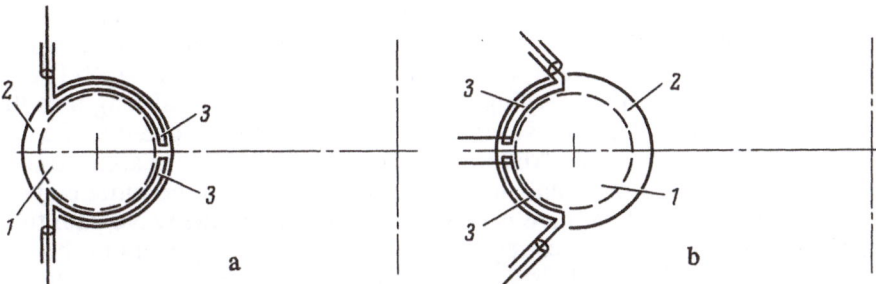

Fig. 5.8. Antenna locations on the high (a) and low field sides (b) for ion cyclotron heating in a tokamak: 1) plasma; 2) limiter shadow; 3) current winding.

must be recovered, and the field calculated inside the opacity region using a simplified method (such as the model of a plane-layered plasma). The properties of the antenna system are modelled in two ways: either using a Fourier expansion of the field in the toroidal angle and following the ray trajectory of each partial wave [166] or by solving the complete wave equation at the plasma edge and using the results in formulating the boundary conditions for the ray trajectories [167]. In a model by Hosea et al. [168], quasilinear distortions in the ion distribution function were also taken into account using a local Fokker–Planck equation. The calculated energy deposition profile was then substituted in the system of equations of the transport model. Calculations using this model generally provided a fair simulation of the results from ICH experiments in PLT [168].

Progress in numerical solutions of the complete system of wave equations in real tokamak geometries [162] has recently risen to the hope of obtaining an exact solution for the energy deposition profile during ICH. Itoh et al. [162] have used the cold plasma approximation and modelled wave absorption by introducing an effective collision frequency. If they are able to obtain a better representation of the rf absorption mechanism, then the energy deposition profile problem may be regarded as completely solved.

5.2. HEATING TECHNIQUES

The range of frequencies used for ICH is usually in the interval between the first and second harmonics of the cyclotron frequency for the bulk plasma component ($\omega_{Bi} < \omega < 2\omega_{Bi}$). For the magnetic fields of thermonuclear systems (2–10 T), this corresponds to frequencies of 30–150 MHz. The problem of developing high-power sources and the corresponding rf circuits in this range does not seem to be especially difficult. If necessary, a steady-state or quasistationary module can be constructed with a power of 5–10 MW, which is quite sufficient for use in a thermonuclear reactor. The greatest technical difficulty is associated with the development of suitable systems for exciting the rf waves in plasmas.

The principal heating schemes use the fast magnetosonic wave. In order to excite these waves in a plasma, it is necessary to create either a toroidal component of the rf magnetic field or a poloidal component of the electric field near the plasma boundary. Antennas in the form of current loops or half loops, which excite the fast wave with a poloidal current, are the most widely used [169–172] (Fig. 5.8). For best matching to the line carrying the rf power, the electrical length of the antenna must be close to 1/4 of the wavelength. The antenna is located in the shadow of the tokamak limiter or in the "divertor shadow" region. For heating in the ion–ion hybrid resonance regime a half-loop antenna mounted on the inner edge of the torus is preferable, since then there is no opacity zone along the wave path from the plasma boundary to the resonance region. One important component of the antenna is an electrostatic shield which "shorts out" the parallel component of the electric field and, therefore, reduces the parasitic slow wave that would be absorbed in the plasma edge. The shield also lowers the plasma density inside the antenna structure.

The basic dimensions of the antenna are determined by the requirements that the parallel (longitudinal) slowing-down spectrum should correspond to optimum penetration of the wave into the plasma and that most of the rf power should be delivered to the plasma. The parallel slowing-down spectrum of a single loop is mainly controlled by the width L of the antenna and extends from low values of $k_\parallel$ to $k_\parallel \sim \pi/L$. The optimum spectral interval corresponds to $k_\parallel > 0.5(n_e/a)^{1/3} \approx 0.1$–$0.3$ cm^{-1}, where n_e is the density in 10^{13} cm^{-3} and a is in cm. This condition sets the maximum width of a single-loop antenna as $L \leq 20$–30 cm.

The impedance of an antenna has an important effect on its efficiency and maximum power. The active component of the impedance is made up of the loss resistance R_l and the load resistance R_L, which must exceed the loss resistance (i.e., $R_L > R_l$) for efficient operation of the antenna. R_l is usually 0.3–$0.5\,\Omega$. The load resistance depends on the dimensions of the current loop and its design and on the wave propagation conditions in the plasma. R_L can be estimated roughly as [51]

$$R_L \approx 0{,}2 B_0 k_{\perp g} d_1^2 L/b \sim 10^{-8} B_0 n_g^{1/2} d_1^2 L/b,$$

which is valid for $k_\parallel(d_1 + d_2) \ll 1$, and $k_{\perp g}(d_1 + d_2) \ll 1$. Here L is the length of the current loop, b is its width, d_1 is the distance from the loop to the return current conductor, d_2 is the distance from the winding to the plasma (all dimensions in cm), $k_{\perp g}$ is the perpendicular wave number at the plasma boundary (cm^{-1}), n_g is the density at the plasma boundary (cm^{-3}), B_0 is the magnetic field (T), and R_L is the resistance (Ω).

Usually the reactive component of the impedance R_I of the antenna is much greater than the active component, so it determines the voltage on the antenna and on the input line. As a rough estimate,

$$R_I \approx 0{,}2 B_0 d_1 L/b.$$

For typical parameters, $R_I = 20$–$50\,\Omega$. At a power of 1 MW this leads to a peak voltage of $U \sim 10$–20 kV. The maximum voltage is essentially the main factor that limits the peak rf power fed to the antenna:

$$P_{max} = (R_L/R_I^2)U_{max}^2.$$

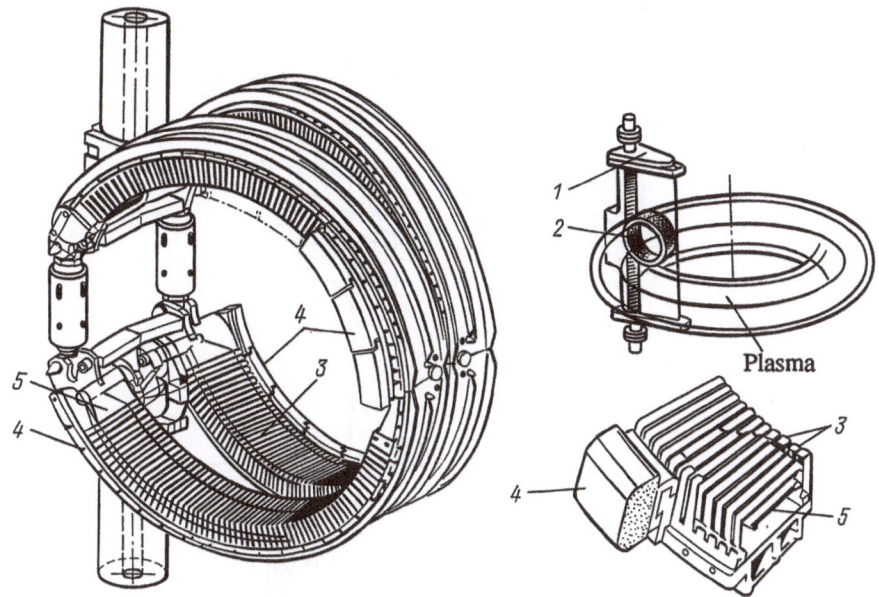

Fig. 5.9. An antenna for the TFR tokamak [170]: 1) tokamak port; 2) antenna; 3) shield; 4) protective limiter; 5) current winding.

Antenna systems have been studied in most detail on the TFR tokamak [170, 173]. Two antennas, each of which consisted of two pairs of current elements (Fig. 5.9), were mounted so that the emission was mainly from the inner edge of the torus (the high magnetic field side). A third antenna was located on the outer edge of the torus. The parameters of these antennas are in fair agreement with calculations employing a numerical code for antennas.

The highest power levels used in ICH experiments have been on the TFR, PLT, and JFT-2 tokamaks [170–173]: up to 0.5–1 MW per antenna, corresponding to a power density of 0.3–1 kW/cm^2. The maximum power delivered to the antenna was limited by breakdowns in the antenna and feeder line. During high-power experiments it was found that radiative losses are the most important limit on heating. It was found that the rapid rise in these losses during rf heating is caused by radiation from atoms of heavy elements that enter from the walls, limiter, and antenna as a result of bombardment by the fast ions that are produced during heating. The impurity influx from near the antenna is especially intense, apparently owing to parasitic heating of the edge plasma. Effective methods of reducing the radiative power loss included the use of light (as opposed to heavy) elements, such as carbon and titanium, for the limiters and antenna components that come into contact with the plasma and the installation of additional limiters near the antenna. In this way it was possible to reduce the impurity radiation losses in PLT, TFR, and JTF-2 to levels considerably below the heating power (at least for pulses lasting up to 0.2 s).

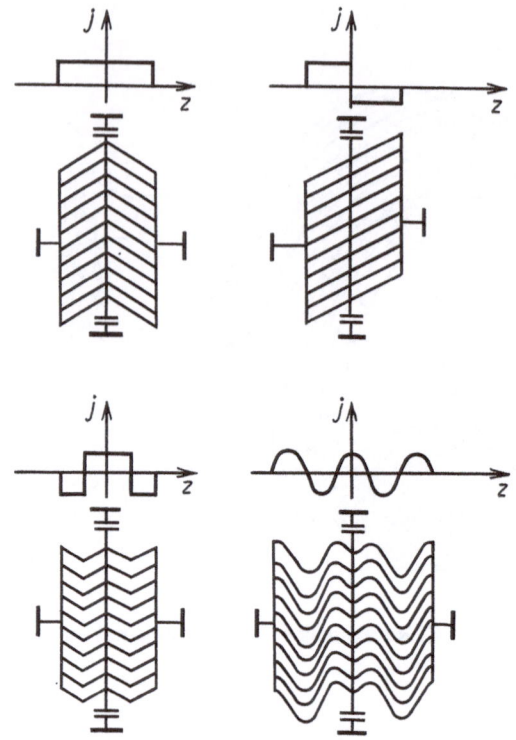

Fig. 5.10. Arrangement of the current surfaces in "fishbone" type antennas [174].

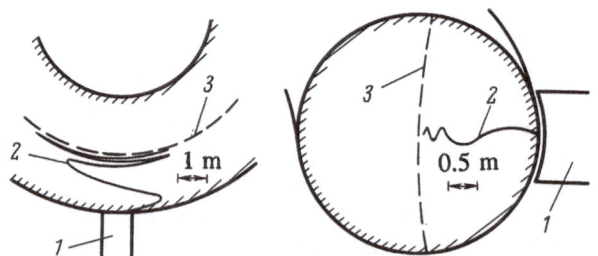

Fig. 5.11. Scheme for launching slow waves in a tokamak [180]: 1) antenna; 2) ray trajectory of the wave; 3) resonance surface.

One proposed alternative to antennas in the form of current-carrying loops or half loops is low-impedance antennas which make it possible to increase the rf input power [174]. A low impedance is obtained by increasing the current conductivity through use of oblique conductors (Fig. 5.10). The field polarization needed for exciting the fast magnetosonic wave is isolated in these antennas by means of an electrostatic shield. Low-impedance antennas were first employed in experiments on the T-10 tokamak. A short 2-MW pulse could be applied to an antenna of this type with an emitting surface of 3500 cm^2.

The question of developing antennas suitable for a thermonuclear reactor has been discussed quite extensively in the literature on ICH. The current elements and loops now employed in experiments are not adequate for use in a reactor. The installation of complicated current elements inside a reactor vessel makes it necessary to solve difficult problems associated with protecting them from bombardment by fast ions and neutrons and ensuring their mechanical durability. Thus, a number of waveguide [168, 175], resonator [176, 177], and slit [177] systems have been proposed for use in reactors. Waveguide systems are the simplest; however, vacuum waveguides in a reactor must be fairly wide ($a > \lambda/2$). For heating at $\omega = 2\omega_{Bi}$ for deuterium, the width must be at least 1.5–2 m for $B_0 = 5$–8 T. Such waveguides can, in principle, be installed, but this will make construction of the magnetic system much more difficult. It has been suggested that the waveguide width be reduced by filling them with a dielectric, using H- or Π-shaped waveguides, or using slit "quasi-waveguide" antennas. Development and computational work on these types of systems are under way.

As noted in Section 5.1, in recent years an ICH scheme in which a slow wave (Bernstein mode) is launched into the plasma and converted into an ion plasma wave has been considered [178–180]. One of the most important advantages of this scheme is the possibility of using a simple waveguide to deliver the wave to the plasma. This advantage originates in the fact that in order to excite the slow mode the electric field of the wave must be parallel to the main magnetic field at the plasma edge. Thus, the launching waveguide must be mounted with its short side in the toroidal direction (Fig. 5.11). The height of the waveguide is limited only by breakdowns and can easily be made compatible with the magnetic field system of a reactor. For example, calculations show that for an input power $P \approx 5$ MW a height of $b \sim 30$ cm would be sufficient. Theoretical analysis confirms the possibility of efficiently exciting the slow wave by means of a single waveguide. In recent initial experiments with this heating scheme, slow waves were excited with the aid of a simple, toroidally oriented loop mounted in the limiter shadow [180]. This sort of loop obviously generates a poloidal rf magnetic field.

5.3. EXPERIMENTAL STUDIES

The first experiments on ICH in toroidal systems were conducted at the end of the 1960s and beginning of the 1970s on the C-stellarator at Princeton [181] and the Uragan stellarator at Kharkov [182]. Ion cyclotron (Alfvén) waves were used and ion temperatures of up to 1 keV at plasma densities on the order of 10^{13} cm^{-3}, were obtained in plasmas with radii of 5–8 cm for input heating powers of up to 10 W/cm^3 in pulses lasting 1–10 ms. The heating scheme had to be changed in order to obtain similar results on larger machines, however, because the excitation of ion cyclotron waves was limited by the existence of an opacity barrier.

For this reason, after the beginning of the 1970's, when ICH experiments were begun on the ST, ATC, TO-1, and TM-1 tokamaks [183–186], they mainly involved the use of fast magnetosonic waves. As noted before (Section 5.1), fast

Table 5.1. Parameters of ICH Experiments in Tokamaks

Machine	Refs.	Experimental parameters				Heating parameters		Heating scheme	Antenna type
		a (cm)	B (T)	$\bar{n}$ (10^{13} cm^{-3})	f (MHz)	P (MW)	τ (ms)		
T-10	[198]	30	3.0	1–3	48	0.5	150	D + H; ωH	"Fishbone"
TFR	[190–194]	25	5.0	5–20	40	2.2	150	D + H; ωH, ω_{ii}	Loop
PLT	[168, 195, 196, 211]	40	1.5–3.0	1–7	25; 42	4.3	200	D + H; ωH D + ^{3}He; ωHe H; 2ωH	Half loops (outer)
Diva	[197]	10	1.2–2.0	1–3	25	0.5	10	D + H;ωH, ω_{ii}	Half loop (outer)
JFT-2	[199–201]	25	1.5	2–6	18	0.5	50	D + H; ωH, ω_{ii}	Half loop (inner)
JFT-2M	[203]	35	1.2	2–5	15	1.6	100	D + H; ω_{ii} H; 2ωH	Half loop (inner)
JFT-2M	[202]	35	1.2	2–5	38	0.6	30	D + H; ωH	Half loops (inner and outer)
JIPP-TII	[204, 205]	17	2.7	2–5	27; 40	0.4	20	D + ^{3}He; ωHe	Loop
TO-2	[208]	8	1.2	1–2	18	0.05	50	H; ωH	Half loop (outer)
JIPP-TIIU	[180]	23	1.5–2.0	1.5	40	0.1	30	D + ^{4}He; slow wave	Longitudinal loop

Table 5.2. Parameters of ICH Experiments in Stellarators

Machine	Refs.	Machine parameters				Heating parameters		Heating scheme	Antenna type
		a (cm)	B (T)	$\bar{n}$ (10^{13} cm^{-3})	f (MHz)	P (MW)	τ (ms)		
Uragan-2	[209]	7	1.5	0.1–2	21	500	2	H; ωH	Loop; slit
L-2	[206,207]	11.5	1.2	1–2	18	150	3	H; ωH	Loop
Heliotron E	[210]	20	2.0	1–3	27	300	15	D + H; ωH, ω_{ii}	Half loop

magnetosonic waves can be efficiently excited in large plasmas since the opacity barrier is narrow for them even when the parallel refractive index is fairly large [when $N_\parallel^3 < N_A^2(0)(\omega a/c)^{-1}$]. It was assumed initially that the main mechanism for absorption of the fast magnetosonic wave would be the resonance at the second harmonic of the ion cyclotron frequency. For this mechanism the optical thickness of the plasma is on the order of $\Gamma \approx v_{Ti}^2 N_A^3 \omega IR/c^3$. In small tokamaks the optical thickness is small ($\Gamma = 0.01$–0.1), so that efficient plasma heating requires that the wave pass repeatedly through the absorption region, which is possible when resonant modes of the plasma column are excited and maintained. This conclusion was confirmed in experiments on the excitation of fast magnetosonic waves in tokamaks. In many tokamaks resonant toroidal modes have been observed, which indicates that multipass absorption is taking place [185, 186]. These resonances manifest themselves as sharp peaks in the amplitude of the wave field and in the load resistance of the antenna. Since the coupling efficiency of an antenna to the plasma is greater near the resonances, methods of maintaining a fixed resonance during the heating process were developed for the experiments on fast wave ICH. The best of the proposed schemes was based on feedback control of the source frequency [187].

In a later stage of experimentation it was found that efficient heating is obtained when a mixture of gases, such as a small additive of hydrogen in deuterium, is used [188–190]. Then substantial absorption can be obtained at the cyclotron resonance of the minority species or at the ion–ion hybrid resonance. These mechanisms make it possible to obtain a large optical thickness Γ, even in small tokamaks, and to realize single-pass heating with strong overlapping of the toroidal modes which practically corresponds to emission into an absorbing medium. Accordingly, in this strongly absorbing regime it is no longer necessary to maintain a fixed resonance in the system.

The parameters of the principal ICH experiments in tokamaks since 1975 are shown in Tables 5.1 and 5.2. The most detailed physical investigations of fast wave ICH have been carried out on the TFR, PLT, and JFT-2 tokamaks. The highest heating powers, 2–4 MW, have been delivered to plasmas in TFR and PLT. Here we shall discuss the experimentally observed heating behavior with fast magnetosonic waves and illustrate it with results from these experiments.

The largest amount of experimental information has been obtained for heating in plasmas containing two ion species. The main gas in these experiments was deuterium, with added hydrogen or, in some cases, ^{3}He. According to the scheme described in Section 5.1, in the minority heating regime (with $n'/n_e < n_{ch}/n_e = N_\parallel v_{Ti}/c$) the minority ions are heated in the neighborhood of their cyclotron resonance surface. The bulk ions are then heated by collisional energy transfer from the minority ions. As the concentration of the minority ions is raised (when $n' > n_{ch}$) a wave conversion region develops in the plasma. This region is separated from the plasma boundary on the outer edge of the torus by an opacity zone (Figs. 5.4 and 5.5). The rf energy is absorbed by electron Landau damping in a region with a large refractive index. Under certain conditions, efficient cyclotron heating of the bulk and minority ions can also occur after conversion (Section 5.1).

This picture of the heating behavior was basically confirmed by experiments. It should first be noted that when hydrogen or ^{3}He is added to bulk deuterium,

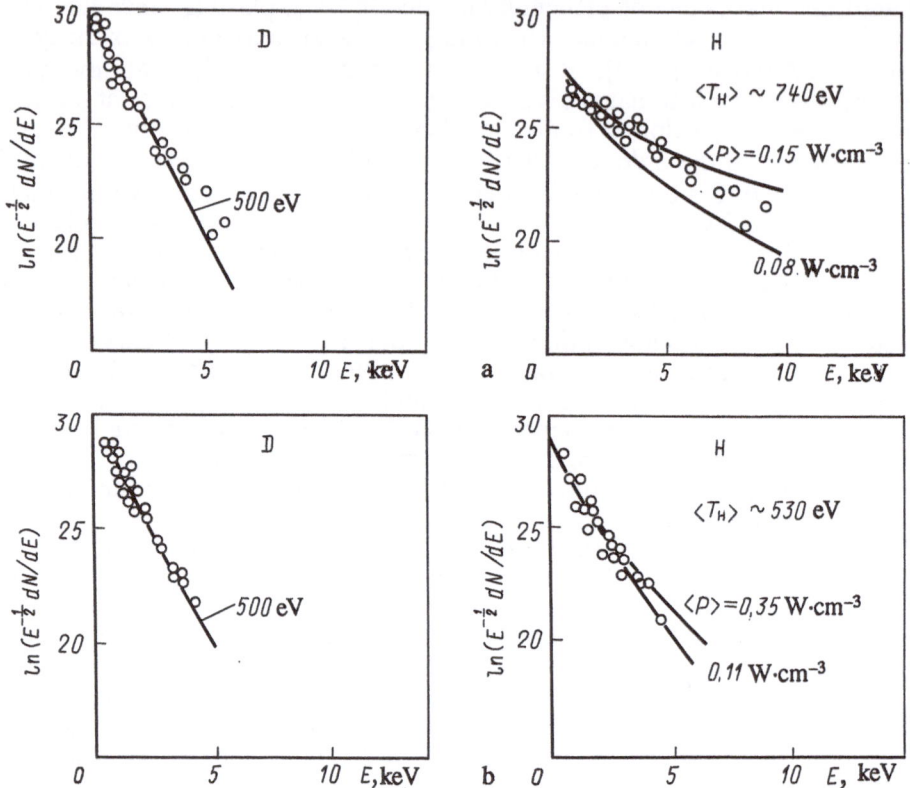

Fig. 5.12. Energy spectra of charge exchange neutrals in experiments on ICH in the JFT-2 tokamak [199] ($B = 1.3$ T, $\bar{n} = 3 \cdot 10^{13}$ cm^{-3}, $P \approx 130$ kW) with $n_H/n_D < 0.03$ (a) and $n/n_D \approx 0.2$ (b).

strong absorption is observed when a cyclotron resonance surface for the minority ions or an ion–ion hybrid resonance surface lies in the central region of the plasma. This is indicated by the overlapping of toroidal modes mentioned above. It has also been demonstrated by measurements of the amplitude of fast magnetosonic waves using magnetic probes [191].

Direct evidence of mode conversion during plasma heating has been provided by measurements of scattering of density fluctuations caused by the pump wave. In TFR slow plasma waves were detected by coherent scattering of infrared radiation from a CO_2 laser and in PLT, by scattering of microwaves ($\lambda \approx 2$ mm). In this way it was possible to determine the spatial distribution of wave amplitudes corresponding to different wave numbers.

When the fraction of minority ions was small, strong heating of these ions was observed when their cyclotron resonance zone lay in the center of the plasma volume. This type of heating greatly increased the fraction of fast ions above that in a Maxwellian distribution. The appearance of a strong "tail" in the distribution func-

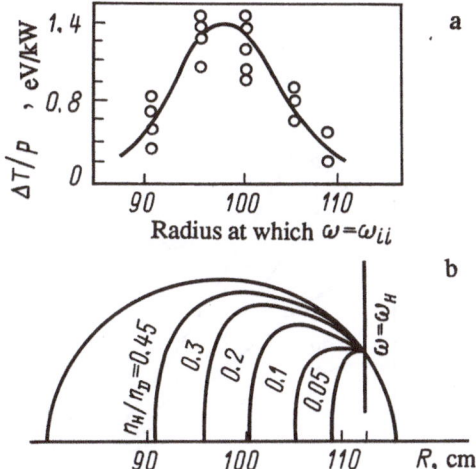

Fig. 5.13. The dependence of the ICH efficiency in TFR on the location of the ion–ion hybrid resonance surface [192] (D + H). The movement of the $\omega = \omega_{ii}(R)$ surface as n_H/n_D is changed is illustrated in (b).

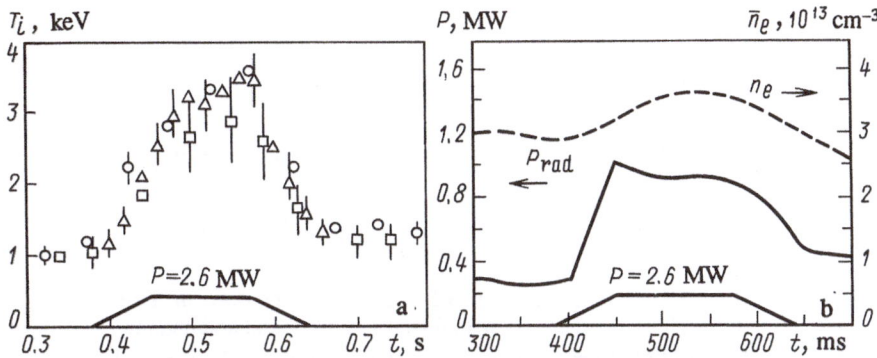

Fig. 5.14. The variation in the plasma parameters during ICH in the PLT tokamak [196] ($B = 3.0$ T; D + ^{3}He; $n_{3He}/n_D < 10\%$; $P = 2.6$ MW): (a) T_i ($\square$ by charge exchange; $\bigcirc$ by Doppler broadening of an Fe XXV line; $\triangle$ by neutrons); (b) $\bar{n}$ and P_{rad}.

tion was first detected in the TM-1 and TO-1 tokamaks [188, 189]. Later it was observed in all experiments on ICH in the minority heating regime. The characteristic form of the energy distribution of hydrogen ions (the added minority species) is shown in Fig. 5.12 with data from experiments on JFT-2. These data are in good agreement with calculations that assume a quasilinear interaction of the resonant ions with the wave (smooth curves in Fig. 5.12). Here the distribution function for the ions of the primary gas (deuterium) differs little from Maxwellian.

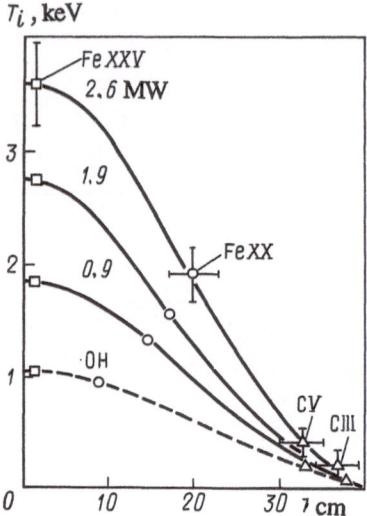

Fig. 5.15. The ion temperature profile during ICH in PLT [196] (D + ^{3}He; n_{3He}/n_D < 10%; $\bar{n}$ ~ $4\cdot10^{13}$ cm^{-3}; data obtained from measurements of broadening of impurity lines).

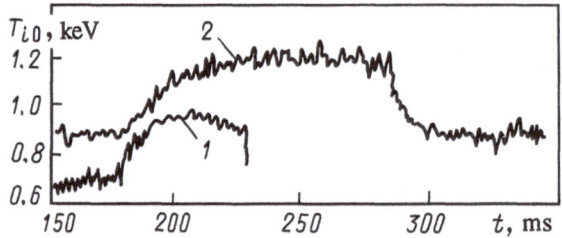

Fig. 5.16. A comparison of data on ion heating in TFR during ICH experiments with different limiter and antenna materials [173] (B = 4.5 T, $\bar{n}$ ~ 10^{14} cm^{-3}; f = 60 MHz; P = 300 kW): 1) Inconel and stainless steel; 2) graphite and titanium.

As the fraction of minority ions is raised, the amount by which they are heated and the distortion in their distribution function at a given heating power are reduced. When the amount of additive is large, the distortions in the distribution functions of both species converge. At the same time, the plasma electrons are heated. The maximum heating effect with a significant fraction of additive ions ($n' > n_{ch}$) is obtained when the hybrid resonance surface is shifted into the plasma center. This is confirmed by plots of the heating efficiency as a function of the magnetic field and minority species fraction (Fig. 5.13). These data show that electron and ion heating take place near the conversion region. It has been proposed that in experiments on

TFR with large minority species fractions ($n'/n_e > 0.3$) heating may have been strongly affected by the absorption of slow plasma waves by heavy impurity ions for which the cyclotron resonance is closer to the conversion region than it is for hydrogen [51]. The energy gained by these ions is transferred to the bulk ions by collisions. Slow wave heating of impurity ions may be especially efficient when the conversion region is close to the resonance zone $\omega = 2\omega_{Bi}'$. In this case an increase in the perpendicular energy may cause escape of impurity ions from the plasma. This effect was apparently observed with Ar XVI ions in TFR [193].

It should be noted that heating with a mixture of ions (when $n' > n_{ch}$) depends on the place at which the fast magnetosonic wave is excited, since an opacity region lies between the plasma edge and the hybrid resonance surface (Fig. 5.5). This region does not prevent the wave from reaching the resonance surface when it is launched from the high field side. Thus, in several tokamaks, differences in heating have been noticed when the waves are excited by antennas mounted on the outer and inner edges of the torus. Fluctuations in the load resistance owing to reflections from the opacity region have been observed with outside launch in the TFR tokamak. Nevertheless, the observed differences in heating with outside and inside antennas were not large; the temperature increases differed by no more than a factor of 1.5. Presumably, the relatively small optical depth of the opacity region ensured almost complete penetration by the wave after a few reflections.

We now consider the plasma heating efficiencies in the PLT and TFR tokamaks. Typical traces of the variations in the plasma parameters during heating pulses are shown in Figs. 5.14 and 5.15. There are clear rises in the ion and electron temperatures during the rf pulse with characteristic times on the order of the corresponding energy confinement times. The ion temperature data were obtained from the energy spectrum of charge exchange neutrals, Doppler broadening of impurity lines, and the intensity of neutron emission.

A density rise was usually observed during heating. Experiments on PLT showed that this is mainly caused by a change in the interaction of the plasma with the wall (recycling) owing to bombardment of the wall by fast ions [196].

An increase in density of impurities (both light and heavy elements) is one of the most important negative effects accompanying ICH. In the first place, an increase in the influx of heavy impurities places limits on the power delivered to the plasma and the duration of the heating pulse. Thus, in TFR the heating efficiency fell sharply 30–40 ms after the beginning of the heating pulse for rf powers of 1–2 MW; the electron temperature fell first and then the ion temperature. The influx of heavy impurities was sharply reduced after graphite limiters were installed in the vessel and near the antennas, and carbon and titanium were used in antenna components that came into contact with the plasma. This led to a reduction in the intensity of heavy impurity line emission in TFR by a factor of 5–10. A similar result was obtained in PLT. This effect can be illustrated by the time variation of the ion temperature for the two arrangements in TFR (Fig. 5.16). It is clear that after the steel and Inconel components were replaced by graphite and titanium it was possible to prevent a drop in the ion temperature late in the discharge. The difficulties associated with the entry of impurities during ICH can apparently be reduced in tokamaks with divertors. This is indicated by a successful ICH experiment in the small Diva tokamak, where deleterious impurity effects were not observed at input powers of up to 10 W/cm^3.

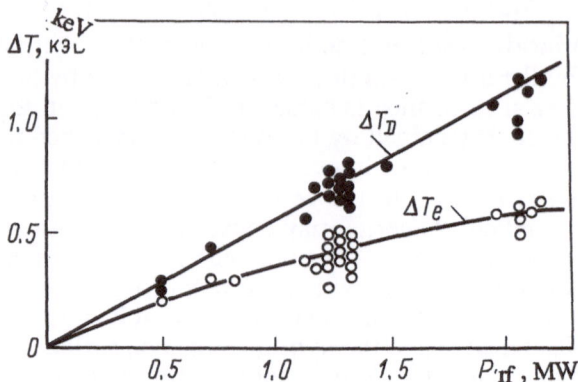

Fig. 5.17. The dependence of the deuterium ion and electron temperature increases on the ICH power in TFR [193] (n_H/n_D = 0.2; $\overline{n} \sim 10^{14}$ cm^{-3}; B = 4.5 T).

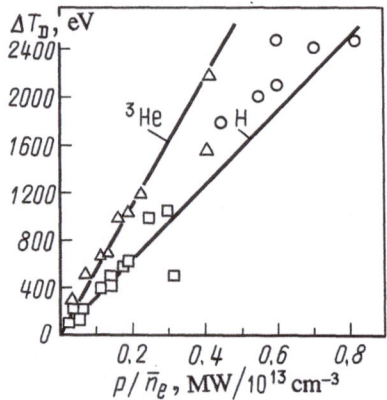

Fig. 5.18. The dependence of the deuterium ion temperature increase on the ICH power in PLT [195]: Δ in the ^{3}He minority heating regime; $\square$ in the H minority heating regime; $\circ$ ion cyclotron heating+neutral beam injection ($P_{ICH} \sim$ 1.6 MW; $P_{NBI} \lesssim$ 2.5 MW).

Table 5.3. Heating Parameters in ICH Experiments on the TFR and PLT Tokamaks

Tokamak	Heating scheme	$\overline{n}$ (10^{13} cm^{-3})	ηh (10^{13} keV/(MW·cm^3))
TFR	D + H; ωii	10	6
PLT	D + ^{3}He; ω_{He}	3	5
PLT	D + H; ω_H	3	2.5
PLT	H; $2\omega_H$	3	3

In most experiments an almost linear dependence of the temperature rise on the rf power delivered to the plasma has been observed (Figs. 5.17 and 5.18). The values of the heating parameter for these experiments are listed in Table 5.3. In recent experiments on PLT the rf power delivered to the plasma has been increased to 4.3 MW. Then the maximum ion temperature reaches 5 keV (at $\bar{n} = 3.5\cdot10^{13}$ cm^{-3}) [211, 246].

Data on the heating efficiency have been obtained in PLT with small minority additives of hydrogen and ^{3}He in bulk deuterium. When the minority species is ^{3}He, the heating efficiency is 1.5–2 times greater than with hydrogen. An analysis shows that this is caused by lower losses of ^{3}He (owing to its smaller cross section for charge exchange in collisions with deuterium atoms) and more rapid transfer of energy from ^{3}He to deuterium (because of a smaller mass ratio). Data have been presented from TFR for hydrogen added to deuterium, when the effect of the ion–ion hybrid resonance is important. The most heating of deuterium is obtained when the minority ion fraction $n'/n_e \sim 20\%$. Then the electron heating is 1.5–2 times less than the ion heating.

On the PLT and, later, on the JFT-2M tokamaks experiments have also been conducted on plasma heating using the fast magnetosonic wave in hydrogen at the second harmonic of the ion cyclotron frequency. It was found that there was significant absorption in regimes with a high ion temperature. The distribution function of the hydrogen ions also deviated significantly from Maxwellian with this type of heating. The increase in the effective ion temperature (determined from the average energy) is comparable to the heating observed in the minority regime (Table 5.3). In JFT-2M substantial electron heating was also noticed. This indicated that conversion into the slow plasma mode had a significant effect. This mode was also detected directly in the 2-mm wave scattering experiments on PLT.

The question of how fast wave heating affects plasma stability and confinement is not conclusively settled. Clear evidence of deterioration in stability has not been noticed experimentally. In most cases numerical models of heating can be matched to experimental data by assuming that the transport coefficient changes by no more than a factor of 1.5 on going from ohmic to rf heating. Usually this change lies within the limits of error in the analysis owing to inaccuracies in the available data on losses associated with impurities and charge exchange. In some cases with high heating powers, a noticeable rise in the energy loss is observed during heating. In particular, in the TFR tokamak the electron energy confinement time fell by a factor of 1.5 during heating. This change was accompanied by a reduction in the electron temperature. It was correlated with a growth in the amplitude of the density fluctuations detected by scattering of infrared CO_2 laser light. In the PLT tokamak the energy confinement time fell by a factor of 2–2.5 at ICH powers of about 1 MW and remained practically constant as the power was raised further (up to 4 MW).

Besides experiments on fast wave heating at the cyclotron resonance for minority additives, at the ion–ion resonance, and at the second harmonic of the ion cyclotron frequency for the bulk species, some recent work deals with heating near the ion cyclotron resonance of the bulk species. The latter results were obtained on small toroidal machines, the L-2 stellarator [206, 207] and the TO-2 tokamak [208]. There a standard antenna was used to launch the fast magnetosonic wave. A heating efficiency comparable to that obtained in other experiments was obtained.

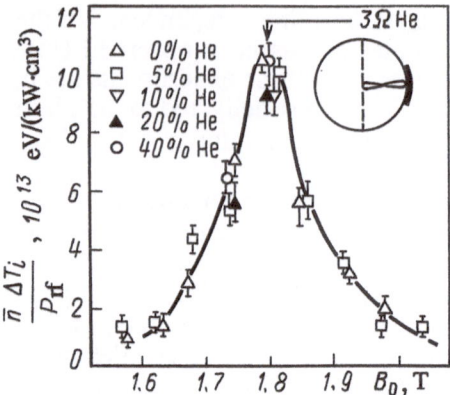

Fig. 5.19. The dependence of the ion heating efficiency in JIPPT-IIU on the magnetic field [180] (H + He; $B = 1.8$ T; $I = 110$ kA; $P = 60\text{–}80$ kW).

The interpretation of the observed heating is still unclear. Since the heating was obtained in small machines with $\overline{n}a < 5 \cdot 10^{14}$ cm^{-2}, it might be supposed that it was caused by excitation of Alfvén waves or by nonlinear effects.

As noted in Section 5.1, ion Bernstein wave heating has recently been discussed as an alternative slow mode for ICH (as opposed to the fast magnetosonic wave). Near the plasma boundary the electric field of this wave is polarized parallel to the toroidal magnetic field [178, 179]. A theoretical analysis has shown that this mode can be efficiently excited in the plasma at frequencies ranging from ω_{Bi} to $2\omega_{Bi}$ for the bulk species. The conditions for propagation depend only weakly on the parallel slowing down for $N_{\parallel} > 2$ and on the plasma density. The perpendicular slowing-down of the wave is very large and it is completely absorbed near the second, third, and higher harmonics of the cyclotron frequencies for additive or impurity ions. One important feature is the perpendicular localization of the ray trajectories owing to the large perpendicular slowing-down (Fig. 5.11).

Calculations of wave excitation have been tested in the small ACT-1 machine [178]. The first experiments on slow wave plasma heating in the JIPP-IIU tokamak have recently been undertaken [180]. These experiments were in hydrogen with a minority helium additive. Heating was intended to take place through absorption of the wave by the minority ions at the third harmonic of the cyclotron frequency. The conditions were chosen such that the $\omega = 3\omega_B'$ surface was at the plasma center. Then the $\omega = \omega_{Bi}$ and $\omega = 2\omega_{Bi}$ resonances for the bulk species (hydrogen) were located outside the plasma. The slow wave was launched with a loop antenna mounted in the equatorial plane along the torus.

A strong heating effect was observed with a neutral particle analyzer that detected charge exchange atoms emitted in the radial direction. The heating efficiency was found to depend strongly on the magnetic field (Fig. 5.19). When the $\omega = 3\omega_{He}$ resonance was located precisely at the center, the heating efficiency (defined in terms of the "perpendicular" temperature of the hydrogen ions) was high, $\overline{n}\Delta T_{i\perp}/P \approx 10^{14}$ eV/(kW·cm^3). The strong dependence of the ion heating on the magnetic field and the high heating efficiency were attributed to localization of the heating.

Contrary to expectations, the heating effect depended only weakly on the amount of minority ions; heating occurred even without them. It was also observed that the rise in the parallel ion temperature (determined with a charge exchange analyzer oriented in the toroidal direction) was significantly less than the rise in the perpendicular temperature. These results can be explained if it is assumed that heating leads directly to an increase in the perpendicular energy of the hydrogen ions, while the rise in the parallel velocity is caused by collisional redistribution of the energy. It was, therefore, concluded that the heating is caused by nonlinear absorption near the surface of half-integer harmonics of the cyclotron frequency of the hydrogen ions, $\omega = (3/2)\omega_H$. This conclusion requires experimental confirmation.

5.4. SUMMARY OF RESEARCH

Research on plasma heating in the ion cyclotron frequency range has led to a high level of understanding of the processes involved in plasma heating with fast magnetosonic waves. Three basic techniques for this type of heating have been proposed: heating through a minority species for which cyclotron resonance conditions are created, ion–ion hybrid resonance heating, and heating at the second harmonic of the ion cyclotron frequency of the bulk ions. Efficient rf antennas with peak power levels of up to 1 MW per antenna have been developed for these heating schemes. The basic features of these techniques have been examined experimentally: the optimum conditions for heating, the evolution of the ion distribution function, and the distribution of the heating energy among the plasma electrons and ions. The greatest difficulty during these experiments has been associated with a rise in the impurity influx caused by bombardment of the vessel walls by fast ions formed during heating and with a sharp increase in the losses through radiation by heavy impurities. The influx of heavy impurities has been effectively limited by using light elements for the limiter and antenna materials. The highest ICH input power levels, 2–4 MW, have been applied to the TFR and PLT tokamaks. The resulting heating was $\Delta T_i \approx 0.5$ keV at $\bar{n}_e \sim 2 \cdot 10^{14}$ cm^{-3} (on TFR) and $\Delta T_i \approx 3$–4 keV at $\bar{n}_e \sim 3 \cdot 10^{13}$ cm^{-3} (on PLT), which corresponds to a heating parameter $\eta_h \sim (2$–6$) \cdot 10^{13}$ eV/(kW·cm^3).

The main problems which must be solved in order to use fast wave ICH in a thermonuclear reactor are the following: first, it is necessary to develop efficient antennas which meet the engineering specifications for the reactor vessel. It would be difficult to meet these specifications using windings and loops similar to those which have been used up to now. The proposed waveguide and slit launching systems appear to be more suitable. These systems, however, require further experimental study. The second problem is associated with the need to restrict the influx of heavy impurities when fast ion "tails" are formed during ICH. It may be more difficult to solve this problem for a reactor than for existing machines. The battle with impurities may be easier in machines with divertors. The third problem, which is common to all heating techniques, concerns the effect of heating on plasma confinement. An effect has been noticed in some experiments. The question is basically whether this effect is specific to the heating techniques being used or whether it is determined by the dependence of transport processes on the plasma parameters.

A scheme for heating using Bernstein (slow plasma) waves has recently been proposed. These waves are launched into the plasma at a frequency between the first and second harmonics of the bulk ions and are absorbed near one of the cyclotron harmonics of minority additive ions. The advantages of this scheme for a reactor are that it is possible to use a simple waveguide launch and the heating is localized. Some doubts arise because of the possible influence of nonlinear effects near the plasma boundaries. The first experiment on the JIPPT-IIU tokamak confirmed that the heating is highly localized and yielded efficiencies 3–4 times higher than for fast wave ICH. The maximum input power was low (on the order of 80 kW) and the heating level was $\Delta T_{i\perp} \approx 0.6$ keV for a plasma density $\bar{n}_e \approx 1.5 \cdot 10^{13}$ cm^{-3}. The prospects for this heating technique must be evaluated in future experiments.

Chapter 6

ALFVÉN WAVE HEATING

6.1. PHYSICS OF ALFVÉN WAVE HEATING

The range of frequencies well below the ion cyclotron frequency seems to be extremely attractive since high-power rf sources have been developed for it [212]. Thus, intensive theoretical studies have been made of the possibility of using electromagnetic waves with $\omega \ll \omega_{Bi}$ for plasma heating [212–214].

A very general idea of wave propagation behavior in plasmas in the Alfvén frequency range $\omega \ll \omega_{Bi}$ can be obtained from Fig. 5.2 as applied to the case $\omega < \omega_{Bi}$. As when $\omega \sim \omega_{Bi}$, two waves can propagate in the Alfvén frequency range: fast magnetosonic and slow (Alfvén) waves. The features of this frequency range are to some extent related to the smallness of the component $g = (\omega/\omega_{Bi})\omega_{pi}^2/\omega_{Bi}^2 \ll \epsilon \simeq 1 + \omega_{pi}^2/\omega_{Bi}^2 + \omega_{pe}^2/\omega_{Be}^2$ of the dielectric tensor. Because it is small everywhere in the plasma except for a small region near the Alfvén resonance $\epsilon = N_{\parallel}^2$, the dispersion equation (2.102) for the fast magnetosonic wave takes the form

$$N_{\perp}^2 = \epsilon - N_{\parallel}^2, \qquad (6.1)$$

where, neglecting the toroidal inhomogeneity of the system in a circular plasma cross section, $N_{\perp}^2 = N_r^2 + c^2 m^2/(\omega^2 r^2)$ and m is the poloidal wave number.

It is clear from Eq. (6.1) that even when m = 0 the fast magnetosonic wave can propagate in a plasma only when $\epsilon > N_{\parallel}^2$. Its cutoffs $N_{\perp}^2 = 0$ are quite close to the resonance point $\epsilon = N_{\parallel}^2$. Using Eq. (2.102), we find that they occur when $N_A^2(r) \equiv \omega_{pi}^2/\omega_{Bi}^2 \approx N_{\parallel}^2 - 1 \pm (\omega/\omega_{Bi})N_A$. Near the resonance $N_A^2 = N_{\parallel}^2 - 1 \approx N_{\parallel}^2$ the fast magnetosonic wave is converted into an Alfvén wave whose dispersion equation [Eq. (2.104)] depends strongly on the parameter $k_{\parallel}^2 v_{Te}^2/\omega^2 = v_{Te}^2/v_{Ar}^2$, where $v_{Ar} = c^2/N_{Ar}^2$ is the Alfvén speed at the resonance. Under typical conditions for present-day experiments, $N_\parallel^2 \gg 1$ at the plasma center and, using Eq. (2.58) and with $v_{Te}^2 \gg v_A^2$, Eq. (2.104) can be written as

$$N_\perp^2 \simeq \frac{2\,\omega_{Bi}|\omega_{Be}|c^2}{\omega^2 v_{Te}^2\,N_\parallel^2}\left(1 + i\sqrt{\pi}\,\frac{c}{v_{Te}N_\parallel}\right)(\epsilon - N_\parallel^2). \qquad (6.2)$$

The imaginary part of this expression describes electron Landau damping. At the plasma edge the condition $k_\parallel^2 v_{Te}^2 \ll \omega^2$ is usually satisfied and instead of Eq. (6.2) we have

$$\epsilon N_\perp^2 = (1 - \omega_{pe}^2/\omega^2)(\epsilon - N_\parallel^2). \qquad (6.3)$$

When $k_\parallel^2 v_{Te}^2 \sim \omega^2$, the slow wave is strongly damped. The fast magnetosonic wave is usually damped only weakly.

Figure 6.1 illustrates this behavior of the perpendicular refractive index $N_\perp^2$ for the fast and slow waves in a plane plasma layer which serves to simulate the plasma near the midplane of a tokamak (neglecting the poloidal magnetic field). The dashed curve denotes the region where Alfvén waves are strongly absorbed.

The main scheme for plasma heating in this frequency range also becomes immediately obvious from Fig. 6.1: heating must be done by an Alfvén wave which is excited in the central region of the plasma through linear conversion of the fast magnetosonic wave. It is useless to launch an Alfvén wave from the vacuum, since its energy will presumably be absorbed at the plasma edge. This scheme for heating establishes the main requirements on the frequency and $N_\parallel$ (parallel slowing-down) of the wave for ensuring an optimum energy deposition profile. First of all, the Alfvén resonance condition

$$N_\parallel^2 \equiv c^2 l^2/(\omega^2 R_0^2) \simeq N_A^2(0), \qquad (6.4)$$

where in a rough model of the toroidicity of the system it is assumed that $k_\parallel = l/R_0$ (with l an integer), must be satisfied in the center of the plasma. Second, the depth of the opacity barrier for the fast magnetosonic wave between the plasma edge and the Alfvén resonance point must be small:

$$(\omega/c)\int_{x_0}^{a} |N_r|\,dx \lesssim 1; \qquad (6.5)$$

otherwise, it is difficult to ensure effective coupling of the antenna to the plasma. For a parabolic density distribution, $\epsilon \simeq N_A(0)(1 - x^2/a^2)$ and Eq. (6.5) can be written in the form

$$(\omega a/2c)N_A(0) \lesssim 1 \qquad (6.6)$$

for $m = 0$ with the aid of Eq. (6.1). When $m \neq 0$, Eq. (6.5) cannot be satisfied. Equation (6.6) does, however, give an order of magnitude estimate of the fre-

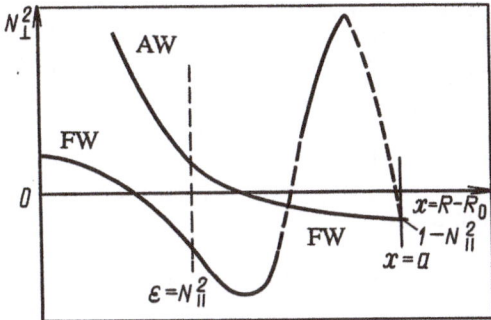

Fig. 6.1. The distribution of the square of the refractive index of the fast and Alfvén waves over the minor cross section of a plasma. (AW denotes the Alfvén wave and FW, the fast magnetosonic wave.)

quency at which the depth of the barrier is a minimum for a given location of the Alfvén resonance.

Equation (6.5), of course, means that the approximation of geometric optics used up to now for the field of the fast magnetosonic wave is no longer valid. The following is a discussion of the basic ideas used in a more rigorous analysis of the wave behavior in the plasma. We note here that although linear wave conversion was explained in terms of geometric optics, the applicability of this approximation to both types of waves is by no means necessary for the existence of conversion. It is sufficient that geometric optics be applicable to the slow wave (or that the slow wave be damped over a short distance). Wave conversion in this case results from a sharp increase in the field of the fast magnetosonic wave near the Alfvén resonance point. In fact, as noted in Section 2, an approximate equation for the fast magnetosonic wave is obtained from the exact wave equation if we set $E_\parallel = 0$. Then, it follows from Eqs. (2.101) that the field component E_x parallel to the vector $\mathbf{k}_\perp$ is given by

$$E_x = ig\,E_y/(\epsilon - N_\parallel^2), \tag{6.7}$$

which implies that this component of the field of the fast magnetosonic wave has a pole at the resonance point. In a more exact treatment (including the finiteness of $E_\parallel$ or perpendicular spatial dispersion), the wave field does not have a singularity and the wave is converted into a slow mode which removes the energy from the resonance.

We note that the residue at the pole in Eq. (6.7) is proportional to the component of the dielectric tensor $\epsilon_{xy} = ig$ associated with the existence in the plasma of a component $\mathbf{j}^{(\perp)}$ of the current induced by the wave that is perpendicular to the field $\mathbf{E}$. Since in the low-frequency limit $\omega/\omega_{Bi} \to 0$ the value of g also approaches zero, it is necessary to include all other mechanisms which might generate a current $\mathbf{j}^{(\perp)}$ perpendicular to $\mathbf{E}$. It is easy to see that the directed velocity $u_{\parallel 0}$ of the electrons caused by the passage of an ohmic current in the plasma will produce such a current. In order to confirm this, the electron contribution to the conductivity tensor

$\sigma_{\alpha\beta}$ can evaluated by shifting to a coordinate system moving with the electrons and using the Lorentz transformation. It is easy to show [213] that including the directed velocity of the electrons in a cylindrical model of the plasma column leads to a change in the component g of the dielectric tensor to $G = g - 2N_{\parallel}c/(\omega qR_0)$, where $q = (r/R_0)B_0/B_{0\vartheta}$ is the stability safety factor (q) of the tokamak.

When the poloidal field is taken into account, $N_{\parallel}$ is related to the wave numbers m and l by the formula $N_{\parallel} = (c/\omega R_0)(m/q + l)$, so that the Alfvén resonance condition can be written in the form

$$\omega = [c\,\omega_{Bi}/(\omega_{pi}R_0)]\,(l + m/q). \tag{6.8}$$

As noted before, in the Alfvén frequency range under realistic conditions the approximation of geometric optics is not applicable and the exact wave equations must be examined. Because these equations are so cumbersome, energy absorption and field excitation by antenna structures can only be studied with the aid of highly simplified equations or numerical methods. A number of such studies [212–214] confirm that efficient delivery of rf power to the plasma center is quite possible both in tokamaks the size of T-10 and in larger machines, such as a tokamak reactor. This conclusion is based on the linear theory. The role of nonlinear effects has not yet been studied adequately, although it has been noted by a number of authors that these effects could certainly appear at the power levels now in use.

6.2. EXPERIMENTAL STUDIES

The first experiments on Alfvén wave heating of toroidal plasmas at frequencies below the ion cyclotron frequency were done on the small stellarators R-0, R-02, Uragan-2, Proto-Cleo, and Heliotron D with low temperatures and low plasma confinement times ($T_e = 10$–100 eV, $\tau_E < 0.1$ ms) [215–221]. These experiments were intended to study the excitation and absorption of Alfvén waves and the distribution of the rf fields in the plasma. They also yielded some information on plasma formation and heating and on current drive. In recent years larger experiments have been started on the TCA, R-05, Pretext, and Tokopol-2 tokamaks [222–226]. The first results on Alfvén heating at parameters typical of tokamaks were obtained on TCA.

Perhaps the most difficult technological problem in realizing Alfvén wave heating is making an efficient antenna system. As noted in Section 6.1, the conditions imposed on the localized Alfvén resonance responsible for conversion and absorption of the waves, which are necessary for heating, can be written as

$$\omega_A = k_{\parallel}v_A = \frac{lB_0}{R\sqrt{4\pi nm_i}}\left(1 + \frac{m}{lq}\right) \approx \frac{lB_0}{R\sqrt{4\pi nm_i}}, \tag{6.9}$$

where l and m are the toroidal and azimuthal mode numbers of the excited wave. The resonance frequency must be less than the ion cyclotron frequency ($\omega_A < \omega_{Bi}$), which leads to the inequality

$$l < R\omega_{pi}/c \simeq 4 \cdot 10^{-6} R\sqrt{n_i/A_i},$$

where R is the major radius (m) and n_i is the ion density (cm^{-3}).

For the actual parameters in toroidal machines, the mode number l obviously cannot be large. This means that in a toroidal plasma a mode must be excited whose parallel wavelength is a large fraction of the circumference of the torus. Thus, in experiments Alfvén waves are excited by antenna systems which extend over the entire length of the toroidal vessel or over a large part of it. Thus, in the R-0, R-02, and R-05 machines, loop antennas made up of 4–8 loops are wound along the entire vessel surface with a step size corresponding to the excitation of specified modes. In these experiments the vessel was made of quartz and the windings were attached to its surface. In the Heliotron D stellarator, waves were launched by a loop whose length was 1/4 of the perimeter of the torus and was mounted inside the metal vacuum vessel outside the separatrix. In the TCA tokamak the launching system was made up of four groups of antennas mounted along the entire torus at equal distances. Each group consists of six windings (three above and three below) made of stainless steel (Fig. 6.2). The antenna structures included an electrostatic shield similar to that employed in experiments on ion cyclotron heating. The phase difference between the antenna groups determined the oscillation mode that was excited. The windings were mounted in the limiter shadow. The surface area of the antenna system was about 16% of the surface area of the vessel.

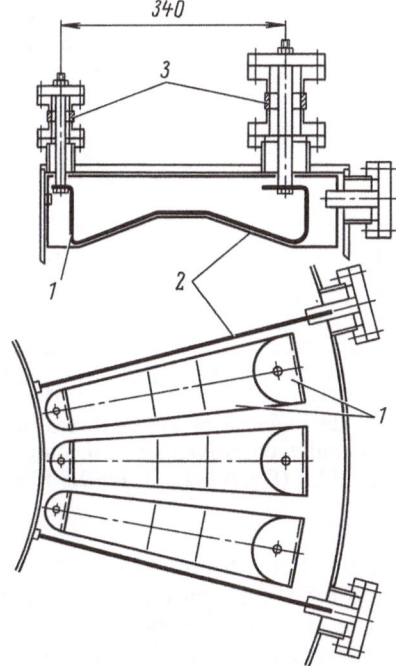

Fig. 6.2. The antenna for Alfvén wave heating on the TCA tokamak [223]: 1) winding; 2) shield; 3) ceramic insulator.

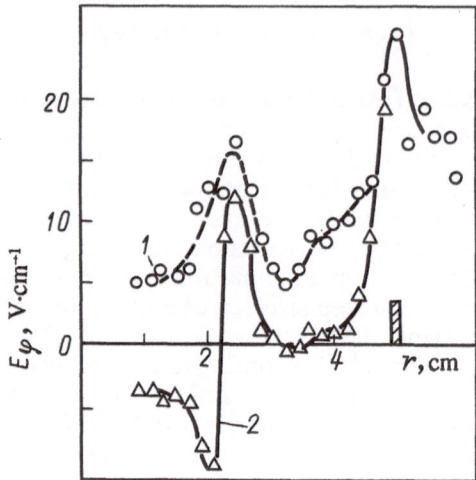

Fig. 6.3. The radial distribution of the azimuthal electric field of the Alfvén wave in the R-0 stellarator [217]: 1) $|E_\varphi|$; 2) Re (E_φ).

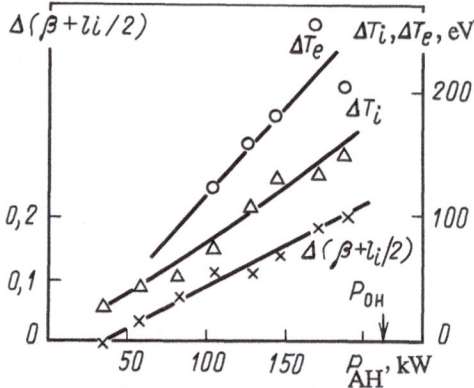

Fig. 6.4. The dependence of the increases in the electron and ion temperatures on Alfvén wave heating power in the TCA tokamak [223] (D_2, $B = 1.5$ T, $f = 2.5$ MHz).

Considerable attention has been devoted in the experiments to studying the excitation of Alfvén waves and their propagation in the plasma. The general picture of wave excitation agrees more or less with theory. Various authors have observed "helical" resonances of Alfvén waves which are isolated from the region where local resonances exist in the plasma (known as the Alfvén continuum) [216, 222]. The resonances also appear distinctly in the active coupling resistance of the antenna system [216, 223]. It should be noted that the coupling resistance in the continuum region depends significantly on the distance between the launching system and the plasma edge, on the location of the local resonance in the plasma, and on the direction of rotation of the rf field (the sign of the ratio of m and l). For paramagnetic rotation (i.e., rotation in the direction of the ion rotation) the coupling resistance is several times greater than for diamagnetic rotation [217]. Typical values of the resistance are 0.1–1 Ω. The field distributions in regimes with low plasma temperatures were observed with the aid of probes. A more interesting result of these measurements was the detection of a sharp peak in the rf fields near the local Alfvén resonance $\omega = \omega_A$ [217, 219, 221]. As an example, Fig. 6.3 shows the measured field distribution in the R-0 stellarator. The location of the observed maximum is close to the calculated position of the local resonance.

Experiments on plasma heating in small stellarators have made it possible to establish some of the characteristics of this heating scheme. Heating is greatest when the Alfvén resonance region is in the central region of the plasma. Then both electron and ion heating are observed. Electron heating, attributable to Landau damping, was stronger in Proto-Cleo and Heliotron D [219–221]. There heating was accompanied by a considerable deterioration in the plasma confinement.

Strong heating was obtained in the R-02 stellarator at very high specific energy inputs (up to 50 W/cm^3) and large rf field amplitudes (up to 10^{-2}) [216]. An ion temperature increase of $\Delta T_i \sim 300$ eV, with considerably less electron heating, was observed under these conditions. In this experiment heating was apparently caused by turbulent processes owing to the interaction of the strong rf fields with the plasma. In experiments on this machine, as well as on the R-0 stellarator [224], MHD instabilities were suppressed and confinement was somewhat improved by applying the rf power. Experiments on R-0 [228] showed that the direction of rotation of the rf field influences how the heating affects confinement. When the field rotates in the direction of the electrons' cyclotron orbits, the discharge contracts and the confinement time increases; with rotation in the opposite direction, the plasma diffusion is enhanced. Strong cyclotron heating of partially ionized impurities, for which the cyclotron frequency was close to the Alfvén frequency, was also observed in R-0. When the cyclotron resonance occurs for the He II and Ne III ions, heating causes removal of these ions from the plasma [224].

We now consider the Alfvén wave heating experiment on the TCA tokamak ($a = 18$ cm, $B = 1$–1.5 T, $I = 140$ kA) in more detail [223]. Powers of up to 200 kW were delivered to the plasma at a frequency of 2.5 MHz in pulses lasting 50 ms. An $l = 2$, $m = 1$ mode was launched using the antenna system described above (Fig. 6.2). The maximum rf input power was on the order of the ohmic heating power. If it was raised further, the discharge disrupted owing to increased MHD activity. It has not yet been possible to eliminate disruptions. The changes in the electron and ion temperatures and in the parameter $\beta + l_i/2$ during heating are illustrated in Fig. 6.4. The maximum electron heating was about 200 eV and the maximum ion heating was about 140 eV. A simultaneous plasma density increase of $\Delta n \sim 2 \cdot 10^{13}$ cm^{-3} was observed. The corresponding normalized heating parameter reached $\eta_h \sim 5 \cdot 10^{13}$ eV/(kW·cm^3). The interpretation of this heating is not completely clear. The electron heating can be explained by Landau damping. The ion heating is partially caused by the growth in the plasma density and, therefore, by increased energy transfer from the electrons. Simulations have shown, however, that part of the power is transferred directly to the ions. The mechanism for this type of heating is still unexplained.

A considerable rise in the radiative power losses was observed during heating. In order to reduce these losses the steel limiters on the TCA tokamak were replaced by graphite ones coated with TiC. This led to a reduction in the influx of heavy impurities and to a sharp drop in the radiative losses from the plasma center (from 2.6 to 0.6 W/cm^3 for heating powers of about 80 kW). In connection with the problems of limiting radiative losses and preventing disruptions, considerable attention has been devoted to measuring the parameters of the edge plasma during Alfvén heating [229]. It has been found that the interaction of the plasma with the antenna leads to strong non-Maxwellian heating of the electrons. This effect can be related to the current flowing from the plasma to the antenna under the influence of the rf fields. This promotes an increase in the impurity influx. In order to limit this effect, it is necessary to mount the antenna deeper in the limiter shadow. In addition, in order to reduce the amount of heavy impurities, an antenna coated with TiC has been made.

6.3. PROSPECTS FOR ALFVÉN WAVE HEATING

We have shown that an efficient heating mechanism exists at frequencies below the ion cyclotron frequency range which involves conversion of the Alfvén wave into a slow mode which is strongly absorbed by electron Landau damping. Experiments have demonstrated the possibility of efficiently delivering energy to a plasma under Alfvén resonance conditions and heating both the electrons and ions. Thus far, the experiments have been on a small scale. The largest experiments have been done on the TCA tokamak with rf input powers of up to 200 kW (close to the ohmic heating power) and have raised the peak electron and ion temperatures by 200 and 140 eV, respectively, at a density on the order of $3 \cdot 10^{13}$ cm^{-3}. This corresponds to $\eta_h \sim 5 \cdot 10^{13}$ $eV/(cm^3 \cdot kW)$.

It should be noted that a whole series of physical questions associated with Alfvén heating are still unanswered. The mechanism for the ion heating observed in many experiments has not been explained. Experiments on TCA revealed a limit on the heating power, owing to the growth of MHD instabilities followed by current disruptions, which has still not been removed. One important problem is the observed effect of Alfvén heating on plasma confinement. In some experiments heating has led to a significant deterioration of confinement, while in others it has led to an improvement. According to one hypothesis, this effect depends on the direction of rotation of the rf waves. The nature of this effect, however, is still not established. Finally, an interaction between the antenna and plasma has been observed which leads to additional impurity influx. In order to reduce the amount of impurities it is necessary to lower the plasma density near the antenna by placing it deeper in the limiter shadow or divertor.

The major technical problem in realizing Alfvén heating is to develop an antenna system which, according to the conditions for exciting the Alfvén resonance, must have a length comparable to the circumference of the torus and cover a large fraction of the vacuum vessel surface (although the rf current leads can be relatively small). This would lead to considerable difficulties in constructing vacuum vessels for large machines or a reactor.

The feasibility of Alfvén heating for thermonuclear systems, therefore, can be evaluated only after the completion of further physical studies, experiments at high heating powers in large machines, and engineering development of antenna systems.

Chapter 7

A COMPARISON OF
PLASMA HEATING TECHNIQUES

7.1. Rf HEATING

We conclude with a brief comparison of rf heating techniques as applied to a thermonuclear reactor. The main advantages and shortcomings of the different methods are compared in Table 7.1. Table 7.2 contains data on the record parameters achieved in heating experiments.

The most developed of the rf techniques at this time is ion cyclotron heating employing absorption of the fast magnetosonic wave in regimes with a minority ion cyclotron resonance, ion–ion hybrid resonance, or second harmonic ion cyclotron resonance (Chapter 5). Up to 4.3 MW of rf power has been delivered to plasmas in PLT (an order of magnitude greater than the ohmic power) with ion heating of up to 4 keV at plasma densities of $(3-4)\cdot 10^{13}$ cm^{-3} [211, 246]. The definite advantages of ion cyclotron heating include its development in terms of physics, with a clear understanding of the main mechanisms for heating, the availability of different heating schemes, and the possibility of optimizing them. It is relatively easy to build high-power sources and energy transmission systems in the ion cyclotron frequency range. The main problem with ICH seems to be the development of efficient antennas that can be accomodated with the engineering design of the vacuum vessel. Difficulties also arise in limiting the influx of impurities caused by bombardment of the walls by the fast ion formed during the heating process.

In terms of the parameters that have been reached, electron cyclotron heating using cyclotron absorption of the ordinary or extraordinary waves by the plasma electrons comes next (Chapter 3). In the record experiment on the T-10 tokamak, a heating power of about 1 MW at a frequency of 83 GHz was delivered to the plasma and electron heating of 3 keV was obtained at a plasma density of $(3-5)\cdot 10^{13}$ cm^{-3} [79]. Studies of ECH have yielded a high level of understanding

$(3–5)\cdot10^{13}$ cm^{-3} [79]. Studies of ECH have yielded a high level of understanding of the mechanisms for wave absorption and plasma heating and aided in the development of optimum techniques for heating under different conditions. The great advantage of this higher frequency range is the possibility of narrowly localizing the region to which heating energy is delivered in the plasma. It is, therefore, possible to ensure heating of the plasma center. Because of this localization, electron cyclotron heating can also be used to control the electron temperature and current profiles and thereby affect the plasma stability. The absence of edge heating also makes it easier to deal with the problems of plasma–wall interaction and impurity influx. It should also be noted that, because the frequency of the rf oscillations is high, there is a greatly reduced possibility of driving instabilities which can affect the plasma confinement and heating. Finally, one clear advantage of electron cyclotron heating is the simplicity with which power can be delivered to the plasma in the vessel because waveguides and antennas in the electron cyclotron frequency range are much smaller than the vessel dimensions. The main problem with electron cyclotron heating is the development of high-power sources at the required wavelengths (1–3 mm). Gyrotrons have been shown to be the best solution to this problem. Getting the required parameters from gyrotron systems, however, still requires some serious development work. In addition, the limits on the plasma density for electron cyclotron heating impose some difficulties. Overcoming these difficulties requires more complicated heating systems which may be somewhat less efficient.

Lower hybrid heating is based on the absorption of strongly slowed-down waves in the neighborhood of the lower hybrid resonance by electron Landau damping or stochastically by fast ions (Chapter 4). The maximum power delivered to a plasma thus far is about 1 MW. In the Petula B tokamak ion heating of 0.5 keV has been obtained at a plasma density of about $5\cdot10^{13}$ cm^{-3} [152, 230], and in the Alcator C tokamak, electron heating of 1.2 keV at a densities on the order of 10^{14} cm^{-3} [129, 231]. The advantages of the lower hybrid frequency range include the relative simplicity of making sources with the required power and systems for delivering the power to the plasma. High-power klystron sources and efficient waveguide technology exist for this frequency range (decimeter wavelengths). Waveguide structures can be used to deliver the waves to the plasma and are easy to bring into the vacuum vessel, since they are much smaller than the vessel dimensions. The physical processes involved in lower hybrid heating are rather complicated and, despite a large amount of research, are not completely understood. In particular, the thresholds for parametric instabilities which lead to strong edge absorption and prevent penetration of the waves to the plasma center are comparatively low in the lower hybrid frequency range. Another major problem is bombardment of the walls by fast ions which releases impurities. This problem arises because lower hybrid heating takes place through the high-energy part of the distribution (as does ion cyclotron heating).

Heating based on the conversion of Alfvén waves has only begun to be used in tokamaks and stellarators relatively recently (Chapter 6), so that there is still not enough experimental data on this technique. The obvious advantage of the Alfvén frequency range is the relative ease of developing efficient high-power sources. The major difficulty is in constructing antennas that are compatible with the vacuum vessel, since the length of the antenna structures must be a large fraction of the circumference of the torus.

Table 7.1. Rf Methods of Plasma Heating: Advantages and Problems

Method	Frequency (MHz)	Advantages	Problems

ECH	$(5\text{–}15)\cdot10^4$	Physical understanding
		
		Localized heating
		T_e profile control and
		  effect on instabilities
		Nonlinear effects weak
		Ease of matching input
		  with vacuum vessel
			Development of
			  high power sources
			
			Limited maximum
			  plasma density

LHH	$(1\text{–}3)\cdot10^3$	Availability of sources
		Relative ease of matching
		  launch system with
		  vacuum vessel
			Refinement of phys-
			  ical understanding
			  of heating
			Overcoming edge
			  heating owing to
			  parametric
			  instabilities
			Wall bombardment by
			  fast ions, impurity
			  influx

ICH	30–100	Physical understanding
		Various heating schemes
		Availability of sources
			Development of
			  antennas com-
			  patible with vessel
			Wall bombardment by
			  fast ions, impurity
			  influx

Alfvén	2–20	Availability of sources	Study of physics
			  of heating
			Development of
			  antennas com-
			  patible with vessel

Table 7.2. The Parameters of Record Experiments on Rf Heating of Plasmas in Tokamaks

Method	Machine	Refs.	P (MW)	τ (s)	$\bar{n}$ (10^{13} cm^{-3})	ΔT_e (keV)	ΔT_i (keV)	η_h (10^{13} eV/(kW·cm^3))
ICH	PLT	[246]	4.3	0.2	3–4	2.5	4	4–5
ECH	T-10	[79]	1.0	0.15	3–5	3	0.3	10
LHH	Petula B	[230]	0.8	0.05	4–7	-	0.6	3–4
	Alcator C	[231]	1.1	0.1	12	1.2	0.8	20
Alfvén	TCA	[229]	0.2	0.05	3	0.2	0.15	5

A central problem with all heating techniques is their effect on plasma stability and anomalous transport of energy and particles. High-power experiments with ion cyclotron, electron cyclotron, and lower hybrid heating have shown that there is a noticeable deterioration of energy confinement (a drop in the confinement time by a factor of 1.5–2). It is not yet clear whether this deterioration is caused by the dependence of transport on the plasma parameters or by particular features of the heating interaction.

7.2. NEUTRAL BEAM INJECTION

The main alternative to rf heating in toroidal machines is the injection of fast neutral atoms [12, 232]. The idea behind this method is simple. A beam of fast neutral atoms produced by charge exchange of a beam of fast positive ions is aimed at the plasma in the vacuum vessel. The energy of the atoms is chosen so that they undergo ionization in the central region of the plasma. The resulting fast ions should be confined magnetically and transfer their energy to the plasma ions and electrons. The best confinement of the resulting ion beam occurs with tangential injection, in which the atom beam is aimed toroidally or at a small angle to the toroidal direction. (At large angles the ions are trapped on wide banana orbits.) The optimum energy of the injected atoms for heating the plasma center is $W = (3-5) \cdot 10^{-18} (\overline{n}a)$, where W is in keV and $\overline{n}a$ is in m^{-2}. In modern machines it is 20–80 keV and is up to 400 keV for the expected parameters of a reactor. The fast ions formed during injection will transfer their energy to the bulk plasma by collisions. When $W > 20T_e$ most of this energy should be transferred to the ions and when $W < 20T_e$, to the electrons. The calculated energy transfer time is much shorter than the Lawson time and with tangential injection it is considerably less than the lifetime of energetic ions in the plasma.

The possibility of heating plasmas with neutral beams is related to the development of high power beam sources, known as injectors. The engineering of neutral beam injectors has already reached a high level [232–234]. Injectors with particle energies of up to 100 keV, total powers of up to 5 MW in pulses as long as several seconds, and efficiencies of about 90% have been built. It should, however, be noted that when the particle energy is raised beyond 100 keV the charge exchange cross section falls rapidly, so that the efficiency of neutral beam injectors decreases.

Up to now a fairly large number of experiments on high power neutral beam heating have been done, especially on the T-11, PLT, PDX, Doublet III, and Asdex tokamaks [20, 235–238] (Table 7.3). The maximum powers delivered to the plasmas were in the range 4–8 MW and ion heating levels of 3–5 keV have been achieved at plasma densities of $(3-10) \cdot 10^{13}$ cm^{-3}. The heating parameter was $(3-5) \cdot 10^{13}$ eV/(kW·cm^3). Neutral beam injection has also recently been employed in experiments on the W-VIIA and Heliotron E stellarators to create and heat current-free plasmas [210, 239].

Relatively high values of β have been obtained by beam heating in a number of tokamaks. A record-high value of $\overline{\beta} = 4.6\%$, the minimum necessary for a thermonuclear reactor, has been obtained in experiments in the Doublet III tokamak [238]. It should be noted that during high-power heating ($P > 1-2$ MW) a reduction of a factor of 2–3 in the energy confinement time of the plasma has been observed, along with a change in its dependence on the plasma parameters (Section 1.3). On the other hand, in tokamaks with divertors conditions have been observed under which the energy confinement time during high-power injection is practically the same as during ohmic heating (the so-called H mode [20]).

Neutral beam injection, therefore, has been extensively and successfully applied to tokamaks and stellarators. Heating powers as high as 5–8 MW, i.e., higher than those in the rf heating experiments, have been used. Plasmas in

Table 7.3. Parameters of the Largest Experiments on Neutral Beam Heating

Machine	P (MW)	W (keV)	$\bar{n}$ (10^{13} cm^{-3})	ΔT_i (keV)	ΔT_e (keV)	η_h (10^{13} eV/(kW·cm^3))
PLT [236]	4	40	2–3	5	3	3–5
PDX [237]	5	40	3–5	3	1.5	3–5
Doublet III [238]	8	75	5–7	5	3	5–7
Asdex [20]	3.5	40	3–5	2	1	2.5–4

tokamaks have been heated to thermonuclear temperatures and maximum values of β have been obtained for the first time with the aid of neutral beam injection. No insurmountable negative effects have been found to be caused by neutral beam injection. The experimentally observed reduction in the energy confinement time may not be specific to beam heating but may be related to the dependence of τ_E on the plasma parameters.

Despite the successes that have been achieved using neutral beam injection in modern experiments, its use in a thermonuclear reactor presents great difficulties. The first is related to a sharp drop in the efficiency with which neutral beams can be created when their energy is raised to the values needed for a reactor ($W > 200$ keV). The possibility of using negative ions in an injector ion source is being considered, but obtaining a large current from such a source is a fairly complicated task. Another problem is related to the large size of injectors with the required power. Installing these injectors near a reactor will cause some loss of the neutron flux (neither neutral atoms nor neutrons can be deflected to the side). Yet another problem is caused by bombardment of the vessel walls by fast ions and neutrals and the related impurity influx. Finally, the construction of reliably operating neutral beam injectors with the high powers needed for a reactor is a complex engineering problem.

7.3. ADIABATIC COMPRESSION

Yet another possibility for plasma heating in tokamaks is adiabatic compression of the plasma by an increasing magnetic field [240, 241]. Two methods of compression have been proposed. The first occurs during a rapid increase in the toroidal magnetic field (minor radius compression). The second method consists of shifting the plasma column along the major radius into a higher magnetic field region (major radius compression). A combination of these two methods, combined compression, is also possible. The required compression time can be estimated from the need to limit diffusive expansion and energy loss during the compression period. This means that the compression time τ_c must be considerably shorter than the energy and particle confinement times. The changes in the density and temperature during adiabatic compression are given by $n_c/n_0 = a$ and $T_c/T_0 = a^{2/3}$, where a is the volume compression coefficient. Compression, therefore, leads to a substantial increase in the stored energy in the plasma compared to its level before compression. The heating power applied to the plasma during adiabatic compression is given by

Thus, the heating efficiency is quite large if rf or neutral beam heating is used before compression. When compression is used, there is a considerable reduction in the auxiliary heating power required before compression to obtain the required temperature. In addition, compression offers other advantages in the use of auxiliary heating since the auxiliary heating can be used before compression, at magnetic fields and plasma densities well below their maximum levels. Thus, electron cyclotron heating can be done at lower frequencies and neutral beam injection will work with lower particle energies before compression.

Experimental studies of minor radius compression have been made on the Tuman-2, Tuman-2A, and Tuman-3 tokamaks [240–243] and major radius compression has been studied in the ATC tokamak and, more recently, on TFTR [244, 245]. Combined compression has been studied on Tuman-3 [243]. These experiments confirmed that heating occurs during compression. Minor radius compression also led to a substantial improvement in the plasma energy confinement and to a reduction in the plasma–wall interaction and impurity influx since the hot plasma region is further from the walls.

These advantages of compression as a heating technique are contrasted with a number of serious technical difficulties. Minor radius compression requires installation of a more complicated system of power supplies and magnetic coils than those in conventional tokamaks in order to permit a rapid rise in the toroidal magnetic field. The design of the vacuum vessel is much more complicated, as it must allow the rapidly rising field to penetrate while sustaining the resulting electrodynamic stresses. Another complication arises from the need to maintain the equilibrium of the plasma column during rapid compression. Finally, the problem of combining a rapid rise in the magnetic field with a superconducting winding should be noted.

A brief review of the advantages and disadvantages of different heating techniques shows that choosing the best method for a thermonuclear reactor requires further research and development. It is quite possible that the optimum heating scheme will be based on a combination of different methods. Presumably, the experiments planned over the next few years on rf heating, neutral beam injection, and adiabatic compression, especially those on the most recent generation of large tokamaks, will make it possible to select the optimum program for bringing a thermonuclear reactor to ignition.

REFERENCES

1. *Heating in Toroidal Plasmas*, Proc. of the Joint Varenna–Grenoble International Symposium, Grenoble (1978), 2 vols.
2. *Heating in Toroidal Plasmas*, Proc. of the 2nd Joint Varenna–Grenoble International Symposium, Como (1980), 2 vols.
3. *Heating in Toroidal Plasmas*, Proc. of the 3rd Joint Varenna–Grenoble International Symposium, Grenoble (1982), 3 vols.
4. *Heating in Toroidal Plasmas*, Proc. of the 4th International Symposium, Rome (1984), 2 vols.
5. V.V. Alikaev, Rf and microwave methods of heating plasmas, in: *Progress in Science and Technology. Plasma Physics Series* [in Russian], Vol. 1, Part 2, VINITI, Moscow (1981), pp. 80–99.
6. E.V. Suvorov, *Izv. Vyssh. Uchebn. Zaved., Ser. Radiofiz.*, **26**, 666 (1983).
7. *Rf Plasma Heating* [in Russian], Materials from an All-Union Conference, Institute of Applied Physics, Academy of Sciences of the USSR, Gorky (1983).
8. L.A. Artsimovich, *Controlled Thermonuclear Reactions* [in Russian], Nauka, Moscow (1963).
9. S.Yu. Luk'yanov, *Hot Plasmas and Controlled Fusion* [in Russian], Nauka, Moscow (1975).
10. K. Miyamoto, *Plasma Physics for Nucl. Fusion*, MIT Press, Cambridge (1976).
11. L.A. Artsimovich, *Nucl. Fusion*, **12**, 215 (1972).
12. V.S. Mukhovatov, in: *Progress in Science and Technology. Plasma Physics Series* [in Russian], Vol. 1, Part 1, VINITI, Moscow (1980), pp. 6–118.
13. H.P. Furth, in: *Fusion*, Vol. 1, Part A, Academic Press, N.Y. (1981), pp. 124–233.
14. S.V. Mirnov, *Physical Processes in Tokamak Plasmas* [in Russian], Énergoatomizdat, Moscow (1983).
15. Equipe TFR, *Nucl. Fusion*, **18**, 647 (1978).

16. S.Yu. Luk'yanov (ed.), *Diagnostics of Thermonuclear Plasmas* [in Russian], Énergoatomizdat, Moscow (1985).
17. M. Murakami, *Nucl. Fusion*, **16**, 347 (1976).
18. V.M. Leonov et al., in: *Plasma Physics and Controlled Nucl. Fusion Research, 1980*, Vol. 1, IAEA, Vienna (1981), pp. 393–403.
19. R.J. Goldston, *Plasma Physics and Controlled Fusion*, **26**, 87 (1984).
20. M. Keilhacker et al., in: *Plasma Physics and Controlled Nucl. Fusion Research, 1984*, Vol. 1, IAEA, Vienna (1985), pp. 71–85.
21. P.C. Efthimion, et al., in: *Plasma Physics and Controlled Nucl. Fusion Research, 1984*, Vol. 1, IAEA, Vienna (1985), pp. 29–44.
22. P.H. Rebut et al., in: *Plasma Physics and Controlled Nucl. Fusion Research, 1984*, Vol. 1, IAEA, Vienna (1985), pp. 11–27.
23. *International Tokamak Reactor, Phase Two A, Part 1*. Report of the International Tokamak Reactor Workshop, Vienna (1983).
24. M.S. Rabinovich, in: *Progress in Science and Technology. Plasma Physics Series* [in Russian], Vol. 2, VINITI, Moscow (1981), pp. 6–79.
25. E.D. Volkov, V.A. Suprunenko, and A.A. Shishkin, *Stellarator* [in Russian], Naukova Dumka, Kiev (1983).
26. L.M. Kovrizhnykh and S.V. Shchepetov, *Nucl. Fusion*, **23**, 859 (1983).
27. V.V. Alikaev et al., *Fiz. Plazmy*, **2**, 390 (1976).
28. N.E. Bogdanov et al., *Fiz. Plazmy*, **10**, 676 (1984).
29. A.I. Akhiezer et al., *Plasma Electrodynamics* [in Russian], Nauka, Moscow (1967).
30. V.L. Ginzburg, *The Propagation of Electromagnetic Waves in Plasmas* [in Russian], Nauka, Moscow (1967).
31. V.L. Ginzburg and A.A. Rukhadze, *Waves in Magnetoactive Plasmas* [in Russian], Nauka, Moscow (1970).
32. T.H. Stix, *The Theory of Plasma Waves*, McGraw-Hill, N.Y. (1962).
33. V.P. Silin, *The Parametric Effect of High-Power Radiation on Plasmas* [in Russian], Nauka, Moscow (1973).
34. R.Z. Sagdeev and M. Rosenbluth (eds.), *The Foundations of Plasma Physics* [Russian translation], Vol. 1, Énergoatomizdat, Moscow (1983).
35. R.Z. Sagdeev and M. Rosenbluth (eds.), *The Foundations of Plasma Physics* [Russian translation], Vol. 2, Énergoatomizdat, Moscow (1984).
36. V.D. Shafranov, in: *Reviews of Plasma Physics*, Vol. 3, M.A. Leontovich (ed.), Consultants Bureau (1967).
37. V.E. Golant and A.D. Piliya, *Usp. Fiz. Nauk*, **104**, 413 (1971).
38. N.S. Erokhin and S.S. Moiseev, in: *Reviews of Plasma Physics*, Vol. 7, M.A. Leontovich (ed.), Consultants Bureau (1979).
39. J. Preinhaelter, *Czech. J. Phys.*, **B25**, 39 (1975).
40. T. Maekawa, S. Tanaka, Y. Hamada, et al., *Phys. Lett.*, **A69**, 414 (1979).
41. I.P. Shkarofsky, *Phys. Fluids*, **9**, 561 (1966).
42. V.E. Golant, *Zh. Tekh. Fiz.*, **41**, 2492 (1971).
43. A. Bernstein and L. Friedland, in: *The Foundations of Plasma Physics*, R. Z. Sagdeev and M. Rosenbluth (eds.) [Russian translation], Vol. 1, Énergoatomizdat, Moscow (1983), pp. 393–443.
44. V.V. Alikaev et al., in: *Rf Plasma Heating* [in Russian], Institute of Applied Physics, Academy of Sciences of the USSR, Gorky (1983), pp. 6–70.

45. A.D. Piliya and V.I. Fedorov, in: *Reviews of Plasma Physics*, Vol. 13, B.B. Kadomtsev (ed.), Consultants Bureau (1987).
46. A.V. Timofeev, *Usp. Fiz. Nauk*, **110**, 329 (1973).
47. V.N. Oraevskii, in: *The Foundations of Plasma Physics*, R.Z. Sagdeev and M. Rosenbluth (eds.) [Russian translation], Vol. 1, Énergoatomizdat, Moscow (1983), pp. 241–278.
48. J. Heading, *Introduction to Phase-Integral Methods* , Wiley (1962).
49. A.D. Frank-Kamenetskii, *Lectures on Plasma Physics* [in Russian], Atomizdat, Moscow (1968).
50. A.D. Piliya and V.I. Fedorov, in: *Rf Plasma Heating* [in Russian], Materials from an All-Union Conference, Institute of Applied Physics, Academy of Sciences of the USSR, Gorky (1983), pp. 281–323.
51. A.V. Longinov and K.N. Stepanov, in: *Rf Plasma Heating* [in Russian], Materials from an All-Union Conference, Institute of Applied Physics, Academy of Sciences of the USSR, Gorky (1983), pp. 105–210.
52. A.A. Vedenov, in: *Reviews of Plasma Physics*, Vol. 3, M. A. Leontovich (ed.), Consultants Bureau (1967).
53. B.B. Kadomtsev, in: *Reviews of Plasma Physics*, Vol. 4, M. A. Leontovich (ed.), Consultants Bureau (1966).
54. A.A. Galeev and R.Z. Sagdeev, in: *Reviews of Plasma Physics*, Vol. 7, M. A. Leontovich (ed.), Consultants Bureau (1979).
55. V.V. Parail, in: *Rf Plasma Heating* [in Russian], Materials from an All-Union Conference, Institute of Applied Physics, Academy of Sciences of the USSR, Gorky (1983), pp. 253–280.
56. T.H. Stix, *Nucl. Fusion*, **15**, 737 (1975).
57. C.F.F. Karney, *Phys. Fluids*, **21**, 1584 (1978).
58. C.F.F. Karney, *Phys. Fluids*, **22**, 2188 (1979).
59. Yu.F. Baranov and V.I. Fedorov, *Proc. of 10th European Conf. on Controlled Fusion and Plasma Phys.*, Moscow (1981), Vol. 1, p. H–13.
60. A. G. Litvak et al., *Nucl. Fusion*, **17**, 659 (1977).
61. Yu. F. Baranov and V. I. Fedorov, *Fiz. Plazmy*, **9**, 677 (1983).
62. Yu.F. Baranov et al., *Proc. of 10th European Conf. on Controlled Fusion and Plasma Phys.*, Moscow (1981), Vol. 1, p. H–8.
63. *Gyrotrons, A Collection of Scientific Papers* [in Russian], Institute of Applied Physics, Academy of Sciences of the USSR, Gorky (1981).
64. A.V. Gaponov et al., *Izv. Vyssh. Uchebn. Zaved., Radiofiz.*, **10**, 1414 (1967).
65. G. Faillon and G. Mourier, in: *Heating in Toroidal Plasmas*, Grenoble (1982), Vol. 3, pp. 1051–1057.
66. K. Felch et al., in: *Heating in Toroidal Plasmas*, Rome (1984), Vol. 2, pp. 1165–1170.
67. *Infrared and Millimeter Waves*, Academic Press, N.Y. (1979).
68. D.G. Bulyginsky et al., in: *Heating in Toroidal Plasmas*, Como (1980), Vol. 1, pp. 187–191; Oxford (1979), Vol. 2, p. 547.
69. C.P. Moeller, in: *Heating in Toroidal Plasmas*, Grenoble (1982), Vol. 3, pp. 1085–1090.
70. M. Thumm et al., in: *Heating in Toroidal Plasmas*, Rome (1984), Vol. 2, pp. 1461–1468.
71. H. Hsuan et al., in: *Heating in Toroidal Plasmas*, Rome (1984), Vol. 2, pp. 809–833.

72. V. Erckmann, in: *Heating in Toroidal Plasmas*, Rome (1984), Vol. 2, pp. 846–852.
73. V.A. Flyagin et al., in: *Heating in Toroidal Plasmas*, Grenoble (1982), Vol. 3, pp. 1059–1065.
74. V.V. Alikaev et al., in: *Problems of Atomic Science and Technology, Series on Thermonucl. Fusion* [in Russian] (1981), p. 3.
75. V.E. Golant et al., *Zh. Tekh. Fiz.*, **42**, 620 (1972).
76. V.V. Alikaev et al., *Pis'ma Zh. Éksp. Teor. Fiz.*, **15**, 41 (1972).
77. V.E. Golant et al., in: *Proc. 6th European Conf. on Controlled Fusion and Plasma Physics*, Moscow (1973), Vol. 1, p. 587.
78. V.V. Alikaev et al., *Fiz. Plazmy*, **9**, 336 (1983).
79. V.V. Alikaev et al., in: *Plasma Physics and Controlled Nucl. Fusion Research, 1984*, Vol. 1, IAEA, Vienna (1985), pp. 419–432.
80. D.G. Bulyginsky et al., in: *Plasma Physics and Controlled Nucl. Fusion Research, 1984*, Vol. 1, IAEA, Vienna (1985), pp. 491–501.
81. R.M. Gilgenbach et al., *Phys. Rev. Lett.*, **44**, 647 (1980).
82. R.C. LaHaye et al., *Nucl. Fusion*, **21**, 1425 (1981).
83. C.P. Moeller et al., *Phys. Fluids*, **25**, 1211 (1982).
84. A.C. Riviere, in: *Heating in Toroidal Plasmas*, Rome (1984), Vol. 2, pp. 795–808.
85. R. Prater et al., in: *Heating in Toroidal Plasmas*, Rome (1984), Vol. 2, pp. 763–778.
86. K. Ohkubo et al., in: *Heating in Toroidal Plasmas*, Grenoble (1982), Vol. 2, pp. 725–731.
87. K. Uo et al., in: *Heating in Toroidal Plasmas*, Grenoble (1982), Vol. 2, pp. 667–676.
88. K. Uo et al., in: *Heating in Toroidal Plasmas*, Rome (1984), Vol. 2, pp. 947–960.
89. R. Wilhelm et al., *Plasma Physics and Controlled Fusion*, **26**, 259 (1984).
90. E.D. Andryukhina et al., in: *Plasma Physics and Controlled Nucl. Fusion Research, 1984*, IAEA, Vienna (1985), Vol. 2, pp. 409–417.
91. D.G. Bulyginsky et al., in: *Controlled Fusion and Plasma Physics* (1983), Aachen, Part 1, pp. 457–460.
92. T.H. Stix, *Phys. Rev. Lett.*, **15**, 878 (1965).
93. G. Tonon et al., Preprint EUR-CEA-FC 1075, Grenoble (1981).
94. P. Lallia, Preprint EUR-CEA-FC 774, Grenoble (1975).
95. F. Troyon and F.W. Perkins, in: *Proc. of the 2nd Topical Conf. on RF Plasma Heating*, Texas Tech. Univ., Lubbock (1974), B4, pp. 1–6.
96. M. Brambilla, *Nucl. Fusion*, **16**, 47 (1976); **19**, 1343 (1979).
97. Yu.F. Baranov and O.N. Shcherbinin, *Fiz. Plazmy*, **3**, 246 (1977).
98. M. Brambilla, in: *Heating in Toroidal Plasmas*, Como (1980), Vol. 1, pp. 285–301.
99. T. Imai, N. Okamoto, and T. Nagashima, Preprint IPPJ–467, Nagoya (1980).
100. T. Nagashima and N. Fujisawa, in: *Heating in Toroidal Plasmas*, Grenoble (1978), Vol. 2, pp. 281–292.
101. V.V. D'yachenko and O.N. Shcherbinin, *Zh. Tekh. Fiz.*, **46**, 1775 (1976).
102. R.W. Motley et al., Preprint PPPL–1651, Princeton (1980).
103. Yu.F. Baranov and O.N. Shcherbinin, *Fiz. Plazmy*, **7**, 646 (1981).
104. Yu.F. Baranov and V.I. Fedorov, *Pis'ma Zh. Tekh. Fiz.*, **4**, 800 (1978).

105. Yu.F. Baranov and V.I. Fedorov, in: *Heating in Toroidal Plasmas*, Como (1980), Vol. 1, pp. 313–316.
106. R. Englade et al., in: *Heating in Toroidal Plasmas*, Como (1980), Vol. 1, pp. 399–404.
107. M. Brambilla and A. Cardinali, in: *Heating in Toroidal Plasmas*, Como (1980), Vol. 1, pp. 405–409.
108. E. Barbato, F. Santini, and A. Taroni, in: *Heating in Toroidal Plasmas*, Como (1980), Vol. 1, pp. 417–425.
109. P.M. Bellan and K.L. Wong, *Phys. Fluids*, **21**, 592 (1978).
110. A.M. Rubenchik, I.Ya. Rybak, and B.I. Sturman, *Zh. Éksp. Teor. Fiz.*, **67**, 1364 (1974).
111. A.M. Rubenchik, *Pis'ma Zh. Tekh. Fiz.*, **2**, 521 (1976).
112. R.L. Berger, L. Chen, and F.W. Perkins, Preprint PPPL–1307, Princeton (1976).
113. R.L. Berger, L. Chen, P.K. Kaw, and F.W. Perkins, *Phys. Fluids*, **20**, 1864 (1977).
114. N.L. Fisch, *Phys. Rev. Lett.*, **41**, 873 (1978).
115. C.F. Karney and N.L. Fisch, *Phys. Fluids*, **22**, 1817 (1979).
116. M. Brambilla and Y.P. Chen, in: *Heating in Toroidal Plasmas*, Grenoble (1982), Vol. 2, pp. 565–569.
117. V.V. Parail and R.V. Shyrugin, in: *Heating in Toroidal Plasmas*, Grenoble (1978), Vol. 2, pp. 475–485.
118. R. Klima, *Plasma Phys.*, **15**, 1031 (1973).
119. S. Tanaka, S. Takamura, and T. Okuda, in: *Heating in Toroidal Plasmas*, Grenoble (1982), Vol. 2, pp. 607–613.
120. J.J. Schuss et al., *Nucl. Fusion*, **21**, 457 (1981).
121. M. Brambilla, in: *Heating in Toroidal Plasmas*, Grenoble (1982), Vol. 2, pp. 505–515.
122. D.L. Grekov, V.E. D'yakov, and K.N. Stepanov, *Pis'ma Zh. Tekh. Fiz.*, **9**, 82 (1983).
123. T.K. Nguyen and D. Moreau, in: *Heating in Toroidal Plasmas*, Grenoble (1982), Vol. 2, pp. 591–597.
124. J.J. Schuss et al., *Bull. Am. Phys. Soc.*, **26**, 1023 (1981).
125. V.K. Decyk, G.J. Morales, and J.M. Dawson, in: *Heating in Toroidal Plasmas*, Como (1980), Vol. 1, pp. 365–375.
126. H. Abe, R. Itatani, and H. Momoto, *Phys. Fluids*, **22**, 1533 (1979).
127. K. Matsuda et al., *Phys. Fluids*, **23**, 1422 (1980).
128. P.T. Bonoli et al., in: *Heating in Toroidal Plasmas*, Rome (1984), Vol. 2, pp. 1311–1318.
129. M. Porkolab et al., in: *Heating in Toroidal Plasmas*, Rome (1984), Vol. 1, pp. 529–545.
130. M. Bernard, in: *Heating in Toroidal Plasmas*, Rome (1984), Vol. 2, pp. 1319–1326.
131. G. Faillon et al., in: *Controlled Fusion and Plasma Physics*, Aachen (1984), part 1, pp. 333–336.
132. A.F. Harvey, *Microwave Engineering*, Academic Press, N. Y. (1963).
133. H. Pacher et al., in: *Heating in Toroidal Plasmas*, Como (1980), Vol. 1, pp. 329–341.
134. I.P. Gladkovskii et al., *Fiz. Plazmy*, **5**, 512 (1979).

135. P. Lallia, *Proc. 2nd. Conf. on RF Plasma Heating*, Texas Tech. Univ., Lubbock (1974), C–1, pp. 1–11.
136. M. El Shaer et al., in: *Heating in Toroidal Plasmas*, Grenoble (1982), Vol. 2, pp. 571–576.
137. R.W. Motley, W.M. Hooke, and G. Anania, *Phys. Rev. Lett.*, **43**, 1799 (1979).
138. K. Uehara and T. Nagashima, in: *Heating in Toroidal Plasmas*, Grenoble (1982), Vol. 2, pp. 485–503.
139. C. Gormezzano et al., in: *Heating in Toroidal Plasmas*, Grenoble (1982), Vol. 3, pp. 1141–1148.
140. M. Porkolab et al., in: *Plasma Physics and Controlled Nucl. Fusion Research, 1982*, IAEA, Vienna (1983), Vol. 1, pp. 227–237.
141. G. Melin et al., in: *Heating in Toroidal Plasmas*, Grenoble (1982), Vol. 3, pp. 1157–1164.
142. INTOR, Phase 2A, Part 1. Report of the International Tokamak Reactor Workshop, Vienna (1983), p. 166.
143. I. P. Gladkovskii et al., *Zh. Tekh. Fiz.*, **43**, 1632 (1973).
144. V. V. Alikaev et al., *Zh. Tekh. Fiz.*, **45**, 523 (1975).
145. M. Porkolab et al., *Phys. Rev. Lett.*, **38**, 230 (1977).
146. C. M. Surko et al., in: *Heating in Toroidal Plasmas*, Como (1980), Vol. 1, pp. 393–398.
147. T. Imai et al., in: *Heating in Toroidal Plasmas*, Como (1980), Vol. 1, pp. 377–384.
148. C. Gormezzano et al., in: *Heating in Toroidal Plasmas*, Grenoble (1982), Vol. 2, pp. 439–453.
149. J.E. Stevens et al., in: *Heating in Toroidal Plasmas*, Grenoble (1982), Vol. 2, pp. 455–468.
150. M. Porkolab et al., in: *Heating in Toroidal Plasmas*, Como (1980), Vol. 1, pp. 355–364.
151. H. Pacher et al., in: *Heating in Toroidal Plasmas*, Como (1981), Vol. 1, pp. 329–341.
152. D. van Houtte et al., in: *Heating in Toroidal Plasmas*, Rome (1984), Vol. 1, pp. 554–570.
153. G. Tonon, *Plasma Physics and Controlled Fusion*, **26**, 145 (1984).
154. V.E. Golant et al., in: *Controlled Fusion and Plasma Physics*, Oxford (1979), p. 46.
155. F. De Marco et al., in: *Heating in Toroidal Plasmas*, Rome (1984), Vol. 1, pp. 546–553.
156. K. Ohkubo et al., in: *Heating in Toroidal Plasmas*, Grenoble (1982), Vol. 2, pp. 543–558.
157. D. Eckhartt et al., in: *Heating in Toroidal Plasmas*, Rome (1984), Vol. 1, pp. 501–512.
158. R.R. Weynants et al., in: *Heating in Toroidal Plasmas*, Rome (1984), Vol. 1, pp. 211–242.
159. V.L. Vdovin, *Nucl. Fusion*, **23**, 1435 (1983).
160. P.L. Colestock and R.J. Kashuba, *Nucl. Fusion*, **23**, 763 (1983).
161. S.C. Chiu and T.K. Mau, *Nucl. Fusion*, **23**, 1613 (1983).
162. S.I. Itoh et al., in: *Heating in Toroidal Plasmas*, Rome (1984), Vol. 1, pp. 407–412.

163. V.L. Vdovin et al., in: *Heating in Toroidal Plasmas*, Como (1980), Vol. 1, pp. 555–559.
164. D.T. Blackfield and J.E. Scharer, *Nucl. Fusion*, **22**, 255 (1982).
165. K.W. Whang and G.J. Morales, *Nucl. Fusion*, **23**, 481 (1983).
166. M. Brambilla and A. Cardinali, *Plasma Phys.*, **24**, 1187 (1982).
167. V.P. Bhatnagar et al., in: *Heating in Toroidal Plasmas*, Como (1980), Vol. 1, pp. 651–567.
168. J. Hosea et al., in: *Heating in Toroidal Plasmas*, Grenoble (1982), Vol. 1, pp. 213–223.
169. O.M. Shvets, V.F. Tarasenko, and S.S. Ovchinnikov, *Zh. Tekh. Fiz.*, **46**, 443 (1966).
170. Equipe TFR, in: *Heating in Toroidal Plasmas*, Grenoble (1982), Vol. 1, pp. 225–241.
171. J. Hosea, *Phys. Rev. Lett.*, **43**, 1802 (1979).
172. JFT–2 Group, in: *Heating in Toroidal Plasmas*, Grenoble (1982), Vol. 3, pp. 1099–1106.
173. J. Adam, *Plasma Physics and Controlled Fusion*, **26**, 165 (1984).
174. A. V. Longinov et al., in: *Proc. 2nd All-Union Conf. on the Engineering Problems of Thermonuclear Reactors* [in Russian], Leningrad (1981), Vol. 1, p. 380.
175. A. V. Longinov, Inventor's Certificate No. 397139; *Byull. Izobret.*, No. 9, (1975); Inventor's Certificate No. 434891; *Byull. Izobret.*, No. 41, (1975).
176. F. W. Perkins and R. T. Kluge, *IEEE Trans. on Plasma Science*, **PS–12**, 161 (1984).
177. O. M. Shvets, S. S. Kalinichenko, et al., *Fiz. Plazmy*, **7**, 485 (1981).
178. M. Ono et al., in: *Heating in Toroidal Plasmas*, Como (1980), Vol. 1, pp. 593–603.
179. S. Puri et al., in: *Heating in Toroidal Plasmas*, Como (1980), Vol. 1, pp. 587–592.
180. T. Watari et al., in: Heating in Toroidal Plasmas, Rome (1984), Vol. 1, pp. 419–426; in: *Plasma Physics and Controlled Nucl. Fusion Research, 1984*, IAEA, Vienna (185), Vol. 1, pp. 523–530.
181. M.A. Rothman et al., *Phys. Fluids*, **12**, 2211 (1969).
182. A. G. Dikin, S. S. Kalinichenko, A. A. Kalmikov, et al., *Plasma Phys.*, **18**, 577 (1976).
183. W. Hooke and J.C. Hosea, *Phys. Rev. Lett.*, **31**, 150 (1973).
184. H. Takahashi, C.C. Daughney, R. Ellis, et al., *Phys. Rev. Lett.*, **39**, 31 (1977).
185. N.V. Ivanov, I.A. Kovan, and E.V. Los', *Pis'ma Zh. Éksp. Teor. Fiz.*, **14**, 212 (1971).
186. V.L. Vdovin, O.A. Zinov'ev, A.A. Ivanov, et al., *Pis'ma Zh. Éksp. Teor. Fiz.*, **14**, 228 (1971).
187. P.L. Colestock, J.C. Hosea, F.C. Jobes, et al., in: *Heating in Toroidal Plasmas*, Grenoble (1978), Vol. 2, pp. 217–228.
188. V.L. Vdovin, V.D. Rusanov, and N.V. Shapotkovskii, *Pis'ma Zh. Éksp. Teor. Fiz.*, **24**, 410 (1976).
189. N.V. Ivanov, I.A. Kovan, and I.A. Sokolov, *Pis'ma Zh. Éksp. Teor. Fiz.*, **24**, 349 (1976).

190. TFR Group, in: *Plasma Physics and Controlled Nucl. Fusion Research, 1976*, IAEA, Vienna (1977), Vol. 3, pp. 39–47.
191. TFR Group, in: *Heating in Toroidal Plasmas*, Grenoble (1978), Vol. 2, pp. 207–216.
192. TFR Group, in: *Controlled Fusion and Plasma Physics*, Oxford (1979), Vol. 2, pp. 355–369.
193. TFR Group, in: *Plasma Physics and Controlled Nucl. Fusion Research, 1982*, IAEA, Vienna (1983), Vol. 2, pp. 17–25.
194. J. Adam, Equipe TFR, in: *Heating in Toroidal Plasmas*, Rome (1984), Vol. 1, pp. 277–290.
195. D. Hwang, M. Bitter, A. Cavallo, et al., in: *Plasma Physics and Controlled Nucl. Fusion Research, 1982*, IAEA, Vienna (1983), Vol. 2, pp. 3–15.
196. J. Hosea et al., in: *Heating in Toroidal Plasmas*, Rome (1984), Vol. 1, pp. 261–276.
197. H. Kimura, K. Odajima, S. Sengoku, et al., *Nucl. Fusion*, **19**, 1499 (1979).
198. A.V. Longinov, I.M. Arkotov, D.V. Berezov, et al., in: *Controlled Fusion and Plasma Physics*, Moscow (1981), Vol. 2, p. 162.
199. JFT–2 Group, in: *Heating in Toroidal Plasmas*, Grenoble (1978), Vol. 1, pp. 259–271.
200. H. Kimura, H. Matsumoto, K. Odajima, et al., in: *Plasma Physics and Controlled Nucl. Fusion Research, 1982*, IAEA, Vienna (1983), Vol. 2, pp. 113–123.
201. M. Matsumoto, *Nucl. Fusion*, **24**, 283 (1983).
202. K. Odajima et al., in: *Heating in Toroidal Plasmas*, Rome (1984), Vol. 1, pp. 243–259.
203. M. Mori et al., in: *Plasma Physics and Controlled Nucl. Fusion Research, 1984*, IAEA, Vienna (1985), Vol. 1, pp. 445–452.
204. T. Amano, T. Fujita, Y. Homada, et al., in: *Plasma Physics and Controlled Nucl. Fusion Research, 1982*, IAEA, Vienna (1983), Vol. 1, pp. 219–226.
205. Y. Miura et al., *Nucl. Fusion*, **24**, 211 (1983).
206. V. A. Batyuk et al., in: *Heating in Toroidal Plasmas*, Grenoble (1982), Vol. 1, pp. 273–283.
207. V. A. Batyuk et al., in: *Controlled Fusion and Plasma Physics*, Aachen (1983), Part 1, pp. 373–376.
208. A. I. Artemenkov et al., *Pis'ma Zh. Éksp. Teor. Fiz.*, **39**, 196 (1984).
209. O. M. Shvets, S. S. Kalinichenko, A. I. Lisoivan, et al., *Pis'ma Zh. Éksp. Teor. Fiz.*, **39**, 196 (1984).
210. K. Uo et al., in: *Plasma Physics and Controlled Nucl. Fusion Research, 1984*, IAEA, Vienna (1985), Vol. 2, pp. 383–395.
211. E. Mazzucato et al., in: *Plasma Physics and Controlled Nucl. Fusion Research, 1984*, IAEA, Vienna (1985), Vol. 1, pp. 433–443.
212. A.G. Elfimov, A.G. Kirov, and V.P. Sidorov, in: *Rf Plasma Heating* [in Russian], Materials from an All-Union Conference, Institute of Applied Physics, Academy of Sciences of the USSR, Gorky (1983), pp. 211–252.
213. K. Appert, J. Vaclavik, and L. Villard, Preprint LRP 187/81, École Fédérale de Lausanne (1981).
214. T. Stix and G. Swanson, in: R.Z. Sagdeev and M. Rosenbluth (eds.), *The Foundations of Plasma Physics* [Russian translation], Vol. 1, Énergoatomizdat, Moscow (1983), pp. 333–364.

215. R.A. Demirkhanov, A.G. Korov, M.A. Stotland, and N.I. Malikh, *Plasma Phys.*, **10**, 444 (1968).
216. R.A. Demirkhanov, A.G. Kirov, S.N. Losovsky, et al., in: *Plasma Physics and Controlled Nucl. Fusion Research, 1976*, IAEA, Vienna (1977), Vol. 3, pp. 31–37.
217. A.G. Kirov, L.F. Ruchko, A.V. Sukachov, et al., in: *Controlled Fusion and Plasma Physics*, Oxford (1980), Vol. 1, p. 18.
218. A.G. Dikij, S.S. Kalinichenko, and Yu.K. Kuznetsov, in: *Plasma Physics and Controlled Nucl. Fusion Research, 1976*, IAEA, Vienna (1977), Vol. 2, pp. 129–243.
219. S.N. Golovato and J.L. Shohet, *Phys. Fluids,* **21**, 1421 (1978).
220. T. Obiki, T. Mutoh, S. Adachi, et al., *Phys. Rev. Lett.,* **39**, 812 (1977).
221. K. Uo et al., in: *Plasma Physics and Controlled Nucl. Fusion Research, 1978*, IAEA, Vienna (1979), Vol. 2, pp. 323–334.
222. A. Chambrier et al., in: *Heating in Toroidal Plasmas*, Grenoble (1982), Vol. 1, pp. 161–172.
223. R. Behn, *Plasma Physics and Controlled Fusion* , **26**, 173 (1984).
224. R.A. Demirkhanov et al., in: *Plasma Physics and Controlled Nucl. Fusion Research, 1982*, IAEA, Vienna (1983), Vol. 2, pp. 91–101.
225. R. D. Bengtson et al., in: *Heating in Toroidal Plasmas*, Grenoble (1982), Vol. 1, pp. 111–160.
226. F. D. Witherspoon et al., in: *Heating in Toroidal Plasmas*, Grenoble (1982), Vol. 1, pp. 197–201.
227. R. A. Demirkhanov et al., *Pis'ma Zh. Éksp. Teor. Fiz.,* **17**, 397 (1973).
228. R. A. Demirkhanov, A. G. Kirov, L. F. Ruchko, and A. V. Skuchaev, *Pis'ma Zh. Éksp. Teor. Fiz.*, **33**, 31 (1981).
229. A. De Chambrier et al., in: *Plasma Physics and Controlled Nucl. Fusion Research, 1984*, IAEA, Vienna (1985), Vol. 1, pp. 531–539.
230. C. Gormezzano et al., in: *Plasma Physics and Controlled Nucl. Fusion Research, 1984*, IAEA, Vienna (1985), Vol. 1, pp. 503–511.
231. M. Porkolab et al., in: *Plasma Physics and Controlled Nucl. Fusion Research, 1984*, IAEA, Vienna (1985), Vol. 1, pp. 463–472.
232. N. N. Semashko et al., *Fast Hydrogen Atom Injectors* [in Russian], Énergoatomizdat, Moscow (1981).
233. N. N. Semashko, in: *Progress in Science and Technology. Plasma Physics Series* [in Russian], Vol. 1, Part 1, VINITI, Moscow (1980), pp. 232–282.
234. W. B. Kunkel, in: *Fusion*, Vol. 1, Part B, Academic Press, N.Y. (1981), p. 103.
235. A. G. Barsukov et al., in: *Plasma Physics and Controlled Nucl. Fusion Research, 1982*, IAEA, Vienna (1983), Vol. 1, pp. 83–84.
236. H. Eubank et al., in: *Plasma Physics and Controlled Nucl. Fusion Research, 1978*, IAEA, Vienna (1979), Vol. 1, pp. 167–198.
237. D. W. Johnson et al., in: *Plasma Physics and Controlled Nucl. Fusion Research, 1982*, IAEA, Vienna (1983), Vol. 1, pp. 9–26.
238. R. Stambaugh et al., in: *Plasma Physics and Controlled Nucl. Fusion Research, 1984*, IAEA, Vienna (1985), Vol. 1, pp. 217–227.
239. W-VIIA Team, G. Cattanei et al., in: *Plasma Physics and Controlled Nucl. Fusion Research, 1984*, IAEA, Vienna (1985), Vol. 2, pp. 371–382.
240. M. G. Kaganskii, *Adiabatic Compression of Tokamak Plasmas* [in Russian], Nauka, Leningrad (1980).

241. V. E. Golant and S. G. Kalmykov, in: *Heating in Toroidal Plasmas*, Como (1980), Vol. 2, pp. 975–983.

242. V. E. Golant, *Plasma Physics and Controlled Fusion,* **26**, 77–86 (1984).

243. G. M. Vorob'ev et al., *Fiz. Plazmy*, **9**, 105–120 (1983)..

244. K. Bol et al., in: *Plasma Physics and Controlled Nucl. Fusion Research, 1974*, IAEA, Vienna (1975), Vol. 1, pp. 83–89.

245. G. D. Tait et al., in: *Plasma Physics and Controlled Nucl. Fusion Research, 1984*, IAEA, Vienna (1985), Vol. 1, pp. 141–154.

246. J. Hosea et al., in: *Controlled Fusion and Plasma Physics*, Budapest (1985), Part 2, pp. 120–123.